Environmental Aspects of Coasts and Islands

edited by

Don Brothwell and Geoffrey Dimbleby

I0894694

Symposia of the
Association for Environmental Archaeology No. 1

BAR International Series 94

1981

B.A.R.

B.A.R., 122 Banbury Road, Oxford OX2 7BP, England

GENERAL EDITORS

A. R. Hands, B.Sc., M.A., D.Phil.
D. R. Walker, M.A.

B.A.R.-S94: "Environmental Aspects of Coasts and Islands".

© The Individual Authors, 1981.

The authors' moral rights under the 1988 UK Copyright, Designs and Patents Act are hereby expressly asserted.

All rights reserved. No part of this work may be copied, reproduced, stored, sold, distributed, scanned, saved in any form of digital format or transmitted in any form digitally, without the written permission of the Publisher.

ISBN 9780860541103 paperback

ISBN 9781407352985 e-book

DOI https://doi.org/10.30861/9780860541103

A catalogue record for this book is available from the British Library

This book is available at www.barpublishing.com

CONTENTS

Please note that additional material is available to download from www.barpublishing.com/additional-downloads.html. The original foldouts have been reduced in size to match the A4 format of this book, the images are therefore not as clear as the original foldouts. Please refer to the original foldouts via the download for the original content.

LIST OF CONTRIBUTORS

M. Bell,
Department of Geography, University of Bristol, University Road,
Bristol BS8 1SS.

J. Bintliff,
School of Studies in Archaeological Sciences, University of Bradford,
Bradford BD7 1DP.

P. Boyd,
Department of Human Environment, Institute of Archaeology, Gordon Square,
London WC1H 0PY.

D. Bramwell,
Fulwood, Baslow Road, Bakewell, Derbyshire DE4 1AA.

D. Brothwell,
Department of Human Environment, Institute of Archaeology, Gordon Square,
London WC1H 0PY.

G. Cowles,
Sub-Department of Ornithology, British Museum (Natural History),
Tring, Herts.

G. Dimbleby,
Department of Human Environment, Institute of Archaeology, Gordon Square,
London WC1H 0PY.

A. Donaldson,
Department of Archaeology, University of Durham, 46 Saddler Street,
Durham DH1 3NU.

D. Gilbertson,
Department of Prehistory and Archaeology, University of Sheffield,
Sheffield S10 2TN.

J. Greig,
Department of Botany, University of Birmingham, Birmingham.

C. Grigson,
Odontological Museum, Royal College of Surgeons of England,
Lincoln's Inn Fields, London WC2A 3PN.

M. Jansma,
Albert Egges van Giffen Instituut voor Prae- en Protohistorie,
University of Amsterdam, Singel 453, Amsterdam, Netherlands.

C. Morris,
Department of Archaeology, University of Durham, 46 Saddler Street,
Durham DH1 3NU.

B. Pickersgill,
Department of Agricultural Botany, University of Reading, Whiteknights, Reading RG6 2AS.

J. Rackham,
Department of Archaeology, University of Durham, 46 Saddler Street, Durham DN1 3NU.

R. Smith,
School of Geography, University of Leeds, Leeds LS2 9JT.

R. Scaife,
Department of Human Environment, Institute of Archaeology, Gordon Square, London WC1H 0PY.

K. Thomas,
Department of Human Environment, Institute of Archaeology, Gordon Square, London WC1H 0PY.

M. Wilkinson,
Department of Prehistory and Archaeology, University of Sheffield, Sheffield S10 2TN.

ENVIRONMENTAL ASPECTS OF COASTS AND ISLANDS: A JUSTIFICATION

Don Brothwell
and Geoffrey Dimbleby

This is the first symposium in what we hope will become a series of established annual meetings to consider topics in environmental archaeology. We would define environmental archaeology as that branch of the study of past human populations which is concerned with evaluating the relationships between the natural environment and Man himself. Broadly viewed, it is human palaeoecology, with all its ramifications, from environmental determinants of early hominid evolution, to Man's impact on the world around him, and even to patterns of disease.

The Association, we hope, will emphasise the breadth, but at the same time the integrity, of the subject, and perhaps humble us all to the fact that not only is it a relatively young part of archaeology but that none of us as yet can claim to be conversant with more than a part of the subject. So our sessions will not only be a means of meeting and talking with one another and hearing something of the work being undertaken in different laboratories, but should also be a continuous process of education for us all, and a means of presenting a united front to archaeology as a whole.

The selection of a topic for this first symposium was not born of any realisation that subsequent meetings might be all at sea if this one was a failure; rather that the maritime area, and the small islands surrounded by sea, are as yet unexplored by the various societies and other bodies which most of us are likely to belong to. Moreover, some of us have had an interest in the biology of islands or coasts in recent years, as the range of contributions demonstrates.

But to return to the title of this meeting, we have been particularly concerned not to be narrow in how we view the coastal zone. The terrestrial ecologists' interest might extent only to the sea edge, the marine biologists' perhaps only to the High Water Mark. Defining a coastal zone on the basis of the distribution of plants such as the marram grass (*Ammophila arenaria*) or salt-marsh grass (*Puccinellia maritima*), we would arrive at a broader band of territory. Ultimately, of course, as environmental archaeologists, we should be influenced in how we view coasts and islands not only by their special ecological characteristics but also by any notable links with human populations. And human activity has embraced, at least on occasion, everything from the cliff top to salt marsh, sand dune and open sea. And this is no modest acreage, if we remember that the total length of coast line in England and Wales alone measures over 2700 miles.

In Semple's classic book on The Influences of Geographic Environment, published in 1911, the boundary between land and sea is recognised as the most important of all, Again, the differences between categories such as coastline, coastal plain, coastal peoples, and so on, are recognised. But essentially, the coast was seen as the 'hem of the sea and land, combining the characters of each' with 'the multifarious influences modifying one another'. From our point of view, two important additional factors are the accessibility of coastal country on the one hand, and of the sea on the other. Caesar in his Bello Gallico noted that the Veneti of Brittany had settlements on the points of capes and promontories, within easy reach of the sea, which at the same time gave them protection from land attack because tidal rises tended to submerge intervening lowlands. By way of contrast, one can think of the arid coastal areas of Peru, where the need to cultivate in pre-Hispanic times led to the specialised invention of mahamaes plots behind the beaches without the need for canal irrigation or a large labour force (Parsons 1968). But the challenge was not always met, and perhaps the unimportance of Ancient Egypt as a maritime nation may have been due to the Egyptians' inability to settle and cope adequately with the dune and marsh environments of the Nile delta.

In a census report of 1861, 'island' was defined as 'a piece of solid land surrounded by water, which affords sufficient vegetation to support one or more sheep'. Surface area varies considerably, of course, from a few square kilometres to over 500,000, and in the latter case much of the land surface is in no way influenced by the sea. So our particular concern at this symposium is the special ecology of the smaller islands. Offshore islands, within sight of the mainland, have often had a succession of human settlers similar to the patterns established on the nearby mainland, although the biological repercussions need not have been the same. In the case of island complexes well separated by sea, settlement by Man has in some areas been very delayed, particularly in the Pacific where pre-European island discovery continued through until at least 500 A.D. (Howells 1973) and perhaps even until the explorations of Magellan, Cook and others.

Although coasts and islands are similar in the extent of their contact with and influence by the sea, there are perhaps some differences of relevance. Islands may have special volcanic or atoll origins. They also tend to have proportionately less fertile alluvial lowlands. As Darwin noted over a century ago, not only may exotic species remain in these isolated habitats, but there may be particularly good evidence of micro-evolutionary diversification—as in his famous Galápagos finches (Lack 1947). At times, possible adaptive changes have been noted and utilised by Man, as in the breeding of the very small Shetland pony from what was clearly a short and shaggy Norse variety of horse (Annandale 1905). Any investigation of the environmental archaeology of past coasts and islands should also keep in mind that human populations have been very varied in these habitats, and care must clearly be taken in estimating population size.

On occupational evidence, pre-ceramic Haldas in coastal Peru may have nurtured the surprising figure of 10,000 people (Rowe 1963), but actual environmental evidence neither refutes nor confirms this. In the case of islands, the same remarkable figure of 10,000 was given in 1841 for the population of a small part of Polynesia, Drummond's Island, with an area of only about twenty square miles (Jenkins 1855). The achievement of relatively high population density in some islands is clearly related to the 'double larder' of land and sea. This demographic characteristic may have a long history in some areas. For instance, might the noticeably common occurrence of prehistoric sites in the Northern Isles indicate relatively high population figures—surviving well on these opportunities for twin harvests?

So far, we have been talking generally, but perhaps now we might view some of the separate aspects of the ecology of coasts and islands in so far as they are relevant to environmental archaeology. In the first place, there is the question of terrestrial topography in relation to the sea. It is not our job here to go into the various coastal classifications (see King 1972), primary and secondary forms, submerging, emerging and eroding coasts, deltaic, volcanic and coral reef coasts, coasts with cliffs, dunes, salt marshes, and mangrove swamps. Differences of this kind and changes through time are clearly linked to what fauna and flora occur and to the exploitation of these areas by earlier human communities. As a non-European example one might perhaps refer to the Sundaic sector of Southeast Asia. This area, significant in human evolutionary terms, has seen considerable sea level oscillations during the past thirty-five thousand years, and although inland changes may have been minor, coastal areas were considerably changed (Dunn and Dunn 1977). Ultimately, there has been nearly a 40% reduction in land mass but nearly a 50% increase in coastline length in this area. Clearly this has relevance to how we view human movements eastward from Sundaland in later Pleistocene times, and would give strength to the view that settlement of the Australian continent by 30,000 years ago would be by raft or boat (Jones 1976). Sundaic cave site assemblages reflect periodic orientations to marine, brackish-estuarine and mangrove resource exploitation, though usually in combination with terrestrial resources as well. Sea beach settlements are also known, but regrettably few middens seem to be intact which are still on land, although perhaps underwater archaeology may in the future reveal other middens now covered by the sea.

Sea-level changes had important consequences, of course, beyond this area. The lower sea-level of the early Postglacial provided lines of migration for plants as well as Man. Not only the North Sea basin, but virtually all the coasts of Europe were important in this respect. A possible west coast route along the Bay of Biscay to Ireland has long interested plant geographers (Matthews 1955).

Similar to the changes in Southeast Asia, one could detail the modifications which have taken place between the contours made by the North Sea in relation to the coasts of south-east England and Holland. But perhaps we might instead mention this area from the point of view of environmental changes which can occur as a result of land reclamation. The Netherlands, in fact, provides a most impressive example of the influence human populations can have in shaping the landscape. Perhaps the origin of this vast reclamation exercise, which has gone on to the present, might be seen in the movement by 300 B.C. of cattle-keeping groups from eastern coastlands onto the Young Marine Clays of northern Friesland and Groningen, which had built up to such an extent since Atlantic (Flandrian Zone II) times that lush salt pastures began to grow on them. But experience with flooding demanded the construction of heightened areas, or terpen. Pliny was clearly not impressed by this sort of thing, noting that the cold inhabitants of these 'tumuli aut tribunalia' warmed themselves by burning 'mud' dried in the wind! But although the country may have been uninviting to the Romans, it nevertheless occupied a position vital to the defence of the western provinces against barbarian tribes, as well as having a position at the mouths of the Rhine and Meuse waterways (Lambert 1971).

In late Roman times and at both ends of the medieval period, there were major marine transgressions (A.D. 250-1600). Terpen were increased in height or abandoned. Dikes still do not appear to have been developed at the time of Viking raids into Frisia, at least according to the Egilssaga, but by A.D. 1000 dike building is thought to have been widely under way (van Giffen 1964), with simple sluice drainage being of great assistance. This was no minor exercise in landscape improvement, and the continuing battle between Man and forces of the sea is demonstrated clearly by the widespread flooding in 1287 between Staveren and the Ems resulting in the estimated death of 50,000 souls, one of a series of such disasters which occurred through time.

Without going into detail, islands can of course show similar changes of topography, soil and vegetation. In Burnt Njal's sad Icelandic saga, the farm at Lithend is described with its productive corn-fields and meadows, and yet by this century the same area is desolate; black sand is being blown over the land, and bogs now occur in some areas of previous agricultural value (Annandale 1905).

So far we have dwelt rather on the general nature of coasts and islands, and on a few examples of landscape change. Although in subsequent papers at this meeting, various other aspects, sites and areas will be considered, perhaps we could give comment on at least some of the special biological characteristics of these environments.

As far as the fauna is concerned, there is obviously much special information to be derived from marine species. The ecology of the shore and offshore waters in relation to Man has received relatively little attention, except for the analysis of mainly fish and molluscs from midden deposits. Molluscs can certainly produce a variety of information as for instance shown by Bailey (1975) on east coast Australian middens. The extent of the biological information depends on the nature of the middens, and Evans (1969) for instance was able to demonstrate changes in relative abundance of various food molluscs through time at a site in the Western Isles. One has to be careful here, of course, because changes in availability may be equally related to changes of sea-level or modifications to the littoral zone, say from a muddy substratum to a rockier shore.

In Japan, Hiroko Koike (1973, 1979) has demonstrated that we are by no means at an end in searching for other kinds of biological data from shell remains, and she has made good use of the daily growth lines of the modern clam (Meretrix lusoria) in estimations of prehistoric seasonal gathering. The application of principal component analysis to data from over 2000 midden clam shells and the identification of shell pairs as a result, suggests that we may hope to extend the range of information in the future on population size, the scatter of paired elements and (with growth line counts) possible seasonal layering.

We see just a little of the importance of this food resource in some past communities by the specialised names which have been handed down through time. In Orkney, relicts of Norse times still remain in such names for invertebrates as sholtie, yug and smurslin. Even the Old Norse name Kvikr, now rendered 'kwick', turns out to refer to lice, so there was more to the Vikings than long boats and pillaging!

In the case of the coastal Maya of the Yucatán peninsula, there is growing evidence that the exploitation and distribution of coastal products formed a potential basis for the support of populations far in excess of the carrying capacity of a maize-orientated, agricultural society (Lange 1971; Ball and Eaton 1972). The case presented sounds very reasonable, and there is no doubt that we have tended to underestimate the potential value of the coastal ecosystem in supporting human communities.

Recent data on the collecting of even one species can provide very impressive evidence of food molluscs available even in some fairly restricted areas. For instance, Jenkins' (1977) figures on cockles gathered in Carmarthen Bay show that yields have been surprisingly high for over forty years, though in decline now.

Before leaving the invertebrates we should of course acknowledge that terrestrial molluscs at coastal and island sites may yield other environmental information, as Penelope Spencer (1975) demonstrated in her study of Cornish and Orkney samples. She mentions, incidentally, the occurrence of razor shells from the neolithic site of Knap of Hower in Orkney, and the fact that these (and associated cockles) indicate a sandy shore habitat not found in the vicinity today. While this could indeed indicate significant ecological changes in the littoral of this area, there is always a danger in using the evidence of a few food animals in this way. Today in Orkney "spoots" (razor shells) are dug for in the sands only at extremely low tides in a restricted period of the year, the foot muscle being a valued delicacy (fried in butter). For this reason they are searched for milés from home base, and brought back alive in their shells.

As far as the vertebrates are concerned we will be hearing something of the work which has been undertaken on fish and birds. However, mention deserves to be made here of a rather neglected category, the marine mammals. Over thirty years ago, Grahame Clark (1946) reviewed the evidence for seal hunting in the "Stone Age" of north-western Europe, but little detailed work has been done on these species since. Whether, as he suggests, the large development of seal hunting in the Scandinavian neolithic results from the "influence exerted by a higher upon a lower form of economy" now seems open to question, particularly if one sees increasing human population pressures as independently stimulating increased if not changing economic activity.

More recently, Fitzhugh (1972) has similarly considered seals in a general evaluation of the environmental archaeology and cultural systems of peoples living in a region of Labrador during the past three thousand years. This economic approach is fine, as far as it goes, but there is really a need for far more effort to be directed towards reconstructing the basic biology of some of these seals, relevant to a full understanding of the relationship between Man and such marine species. Some recently excavated samples are quite large and deserve more effort than basic reporting. Over 230 seal bones and teeth were for instance excavated at the Scillonian site of Nor Nour during just two seasons (Turk 1969-1970). There were both young and adult specimens, with a puzzling excess of skull remains. Considering the relatively small numbers of seals in the Cornwall-Scillies area today (Hewer 1974), we may well question whether the common occurrence of seals at this site indicates a phase of intensive sealing or the alternative possibility of greater numbers of seals at this earlier date.

Incidentally, studies on island seals demonstrate without doubt that it is not good to be raised in a harem! Apparently this type of life more easily leads to inadequate mother/child bonding, with greater risk of starvation and infection as a result (Le Boeuf and Briggs 1977; Anderson, Baker, Prime and Baird 1979), One can only speculate at present as to the role human groups may have played in keeping seal numbers down in some areas, and as a result perhaps ensuring healthier populations. More detailed skeletal studies might eventually tell us far more about age group compositions and perhaps even evidence of disease in such remains. This link between Man and seal reaches Freudian dimensions in the Orkney stories of selkie folk—seals in human form. These male marine gallants had a predilection for married women, we are told, and the result was sometimes children with webbed hands and feet (Dennison 1961). Folklore has it that the cutting of this webbing in the offspring resulted in the growth of horny crusts on the palms and soles, and it is surely a unique ecological story that as a result of microevolution within the Orcadians, the identical clinical condition known as tylosis has now been shown to be noticeably in evidence (Muir 1978).

But to return to ground level! We shall be hearing in following papers something of early plant communities along the Peruvian coast and in the Orkney and Scilly Isles, as well as the relevance of some intertidal species to Man, so our comments here can again be purely introductory. In general terms, one can say that by coast and island studies one can hope to extend the variety of information on plant species and communities of relevance in archaeological reconstructions. For instance in British dune systems, there are about four hundred established species of vascular plants. There are also many additional species more recently introduced, either associated with forestry or as elements derived from the weed flora of agricultural land (Salisbury 1952; Ranwell 1972). The extent of dunes and their vegetation can of course change noticeably, as seen for instance in the historical changes to the sands of Forvie in Scotland (Gimingham 1964).

It is relevant here to remember that some of our important plant cultivars came from the coastal region (Dimbleby 1978). Examples would include cabbage variants, beet, fennel, seakale and asparagus, all cultivated as leaf-vegetables, probably at first in the Mediterranean area. Additional species would include the coconut palm, which is 'designed' for sea dispersal and is clearly a coastal plant.

Maritime and sub-maritime plant communities are also usually subject to an oceanic type of climatic regime. The sea usually has a moderating effect upon coastal temperatures, but this "oceanicity" has been a changing influence through time and space. Not only temperature and humidity are involved here, but also the long term effect of winds, and this seems to have been a major factor in the burial by dune activity of various human coastal settlements in medieval Europe. In some areas at least, coastal margins and uninhabited island communities have been less affected by Man than more inland areas. On the other hand, dense populations of sea birds can have an equally devastating effect, though in far more restricted localities. Areas influenced by tides or occasional flooding or sea spray are likely to have a significant proportion of halophytes, able to cope with the high salinities.

More tropical maritime plant communities provide further evidence of variation. Perhaps the most exotic examples are the mangrove swamp forests in the tidal zones of low, humid, alluvial coasts. Evidence of mangroves along the Pacific littoral of Ecuador appears to extend back to 6500 B. C. and Sarma (1976) claims that the mollusc *Anadara tuberculosa* is habitat specific to mangrove swamps and can thus be used as a reliable indicator. These lowland areas, from Ecuador through to as far north as Guatemala show a paucity of people, and although poor land quality is one factor in some areas, the major factor may be that they are excellent habitats for certain forms of diseases, especially gastro-intestinal organisms, yaws and malaria. In lowland Colombia, early Amerindian settlement certainly occurred in the south around the Tumaco-Esmeraldas area, and there are numerous mounds (tolas) and other sites even on old beach remnants within the mangrove and freshwater swamps near the coasts, but the peoples had vanished by the time the Spanish arrived (West 1967). This tantalising who-dun-it is also considered by Coe and Flannery (1967), working in south coastal Ecuador. In 1579, the Spaniard Estrada records that although salt-making was sometimes carried out near the coast, few Amerindians lived there. Disease, particularly malaria, seems to have been the reason for the depressed population, which seems to have plunged by late Classic times (*c.* 750-900 A. D.). Could it be, ask Coe and Flannery, that malaria or another devastating disease hit these communities long before the Spaniards arrived? So we are back yet again to the old problem of possible corridors of contact across the Pacific!

The exploitation by Man of coastal and island environments from a botanical point of view rather follows the pattern set for other more fertile land areas. In some regions agriculture can extend close to coastal margins. Even an island group in the more severe climate of the north, as in the Shetland Isles, may nevertheless show the basic pattern of change resulting from land use (Spence 1979). From the pollen evidence at Murraster in Shetland, we can see that even in such a wind-swept and salty environment, the same pattern of evolution to arable farming occurs, initially with loss of trees and shrubs, the increase in heather (*Calluna*) and finally the appearance of ribwort (*Plantago lanceolata*) (Johnston 1975). In the case of the Isles of Scilly, not only have we further plant evidence showing change, but there are submerged field systems available for study as well (Fowler and Thomas 1979).

Compared with these general agricultural practices, the specialised exploitation of coast and island resources seems to have been somewhat restricted. Excluding seaweed, some other plants have obviously been tested out by Man for their possible economic or medicinal value. The Amerindians of more southern coastal California used part of sea-blite (*Suaeda californica*) from salt marshes as a rich black dye, and the seeds for flour making (Balls 1962). In the island of Curaçao, and in contrast to some neighbouring island groups, more than 100 species of local plants are employed for self-midication. Ominously, the incidence rate for oesophageal cancer is one of the highest in the world (20.9 per 100,000), and it seems likely that this is related to the use in the islands of one or other of their special indigenous plants (Norton 1968).

Bibliography

Anderson, S. S., Baker, J. R., Prime, J. H. and Baird, A. 1980. Mortality in grey seal pups: incidence and causes. J. Zool. In press.

Annandale, N. 1905. The Faroes and Iceland: Studies in Island Life. Clarendon, Oxford.

Bailey, G. N. 1975. The role of molluscs in coastal economies; the results of analysis in Australia. J. Arch. Sci. 2, 45-62.

Ball, J. W. and Eaton, J. D. 1972. Marine resources and the prehistoric lowland Maya: a comment. Amer. Anthrop. 74, 772-6.

Balls, E. K. 1962. Early Uses of California Plants. University of California Press, Berkeley.

Clark, J. G. D., 1946. Seal hunting in the Stone Age of North-Western Europe: a study in economic prehistory. Proc. Prehist. Soc. 12, 12-48.

Coe, M. D. and Flannery, K. V. 1967. Early Cultures and Human Ecology in South Coastal Guatemala. Smithsonian Press, Washington.

Dennison, W. T. 1961. Orkney Folklore and Traditions. Herald Press, Kirkwall.

Dimbleby, G. W. 1978. Plants and Archaeology. The Archaeology of the Soil. Paladin, London.

Dunn, F. L. and Dunn, D. F. 1977. Maritime adaptations and exploitations of marine resources in Sundaic Southeast Asian prehistory. pp. 1-28. In: G-J. Bartstra, W. A. Casparie and I. C. Glover (eds.) Modern Quaternary Research in Southeast Asia. Baljema, Rotterdam.

Evans, J. G. 1969. The exploitation of molluscs. Pp. 479-84. In: P. J. Ucko and G. W. Dimbleby (eds.) The Domestication and Exploitation of Plants and Animals. Duckworth, London.

Fitzhugh, W. W. 1972. Environmental Archaeology and Cultural Systems in Hamilton Inlet, Labrador. A Survey of the Central Labrador Coast from 3000 B.C. to the Present. Smithsonian Contributions to Anthropology, No. 16. 299 pp. Smithsonian Institution Press, Washington.

Fitzhugh, W. (ed.) 1975. Prehistoric Maritime Adaptations of the Circumpolar Zone. Mouton, The Hague.

Fowler, P. and Thomas, C. 1979. Lyonesse revisited: the early walls of Scilly. Antiquity, 53, 175-89.

Gifffen, A. E. van. 1964. De ouderdom onze dijken, Tidjschr. K. ned. aardrijksk Genoot., 81, 271-86. Quoted by Lambert, 1971.

Gimingham, C. H. 1964. The maritime zone. Pp. 67-142. In: J. H. Burnett (ed.) The Vegetation of Scotland. Oliver and Boyd, Edinburgh.

Hewer, H. R. 1974. British Seals. Collins, London.

Howells, W. 1973. The Pacific Islanders. Weidenfeld, London.

Jenkins, J. S. 1855. United States Exploring Squadron (1838-42). Quoted by Semple, 1911.

Jenkins, J. G. 1977. Cockles and mussels: aspects of shellfish-gathering in Wales. Folk Life, 15, 81-95.

Johnston, J. L. 1974. Shetland habitats, an outline ecological framework. Pp. 33-49. In: R. Goodier (ed.) The Natural Environment of Shetland. Nature Conservancy Council, Edinburgh.

Jones, R., 1976. Tasmania: aquatic machines and off-shore islands. Pp. 285-63. In: G. de G. Sieveking, I. H. Longworth and K. E. Wilson (eds.) Problems in Economic and Social Archaeology. Duckworth, London.

King, C. A. M. 1972. Beaches and Coasts. Arnold, London.

Koike, H. 1973. Daily growth lines of the Clam, Meretrix lusoria—a basic study for the estimation of prehistoric seasonal gathering. J. Anthrop. Soc. Nippon, 81, 122-38.

Koike, H. 1979. Seasonal dating and the valve-pairing technique in shell-midden analysis. J. Arch. Sci. 6, 63-74.

Lack, D. 1947. Darwin's Finches. Cambridge University Press.

Lange, F. W. 1971. Marine resources: a viable subsistence alternative for the prehistoric lowland Maya. Amer. Anthrop. 73, 619-39.

Lambert, A. M. 1971. The Making of the Dutch Landscape. An Historical Geography of the Netherlands. Seminar Press, London.

Le Boeuf, B. J. and Briggs, K. T. 1977. The cost of living in a seal harem. Mammalia, 41, 167-95.

Matthews, J. R. 1955. Origin and Distribution of the British Flora. Hutchinson's University Library, London.

Morton, J. F. 1968. A survey of medicinal plants of Curaçao. Econ. Bot. 22, 87-102.

Muir, V. M. L. 1978. Tylosis in the Orkney Islands. J. Biosoc. Sci. 10, 1-6.

Parsons, J. R. 1968. The archaeological significance of mahamaes cultivation on the coast of Peru. Amer. Antiq. 33, 80-5.

Ranwell, D. S. 1972. Ecology of Salt Marshes and Sand Dunes. Chapman and Hall, London.

Rowe, J. H. 1963. Urban settlements in Ancient Peru. Nawpa Pacha, 1, 1-28. Institute of Andean Studies, Berkeley.

Salisbury, E. J. 1952. Downs and Dunes. Bell, London.

Sarma, A. V. N. 1976. Ecuadorian mangroves: prehistoric and recent. some observations. El Dorado, 1, 88-95.

Semple, E. C. 1911. Influences of Geographic Environment on the Basis of Ratzel's System of Antorpo-Geography. Constable, London.

Spence, D. 1979. Shetland's Living Landscape. A Study in Island Plant Ecology. Thule Press, Sandwick.

Spencer, P. J. 1975. Habitat change in coastal sand-dune areas: the molluscan evidence. Pp. 96-103. In: J. G. Evans, S. Limbrey and H. Cleere (eds.) The Effect of Man on the Landscape: The Highland Zone. CBA Research Rep. 11, London.

Turk, F. A. 1969-1970. Faunal remains. Nor Nour excavations. On file as A. M. Lab. Reports 1537 and 1538. Department of Environment, London.

West, R. C. 1957. The Pacific Lowlands of Colombia. A Negroid Area of the American Tropics. Louisiana State University Press, Baton Rouge.

ARCHAEOLOGY AND THE HOLOCENE EVOLUTION OF COASTAL PLAINS IN THE AEGEAN AND CIRCUM-MEDITERRANEAN

John Bintliff

My introduction to environmental archaeology came through field-work with Eric Higgs in Israel and Italy, where I helped on the study of numerous 'site catchment territories', using the Catchment Analysis approach to understand the interaction of past communities with their natural environment (Higgs and Vita-Finzi, 1970).[1] Even in the early stages of the Cambridge History of Agriculture Project it was being realised that the present day landscape might differ significantly from the ancient or prehistoric one, and that the degree and localisation of such changes must be estimated to make Catchment Analysis a valid exercise. However, it is certainly true of several published Catchment papers, that site and even regional palaeoecological interpretations would be very different had landscape change been allowed for.[2] Whilst in Israel I assisted Vita-Finzi in his research on the effects of recent geomorphic changes on site territory potential (Webley, 1972).[3] Soon afterwards I began my own research in Greece, which was to concentrate on the interrelation of the natural environment with successive periods of settlement and land use from Neolithic up to modern times. A number of regions was selected, in each of which all the published archaeological sites were studied, together with the regional physical environment, in the field and from the literature (Bintliff, 1977).[4]

Already in the first region, the Argolid peninsula, there was an embarrassing absence of previous data on Quaternary Landscape change. The approach adopted was to take the numerous sites being discovered by an American archaeological site survey and relate them to local soil types; apart from Tertiary and older deposits, favoured by settlement. I was very struck by the virtual absence of sites from a red, poorly sorted, supposed Quaternary formation, and the complete absence from a brown, arguably Holocene alluvial silt. This led me back to study of Vita-Finzi's book, The Mediterranean Valleys (Cambridge, 1969), where he put forward the general theory of two fundamental sedimentary episodes in Mediterranean valleys with those characteristics, the Older and Younger Fill. Study of the Argolid sediments showed them to correspond very closely to those described there, and indeed I dated them by associated artefacts to Würm glacial and Medieval times respectively, as predicted by Vita-Finzi.

The Older Fill (Figure 1) is a poorly sorted sediment, primarily colluvial but grading into alluvial facies, a jumbled formation with everything from angular scree to fine clay, reddish in hue, and strongly represented in both piedmont, as here, and full plain topography. The red hues seem to suggest a highly oxidised environment, whilst the steep angle of the formation at the coast, visibly descending well below present sea-level, argues for its deposition

Fig. 1

Fig. 2

during glacial low sea-level. It is suggested that the early to mid-glacials in the Mediterranean saw more pronounced rainfall and greater frost shattering in order to produce these exceedingly widespread deposits; they are often tens of metres deep and probably make up most Mediterranean valley and plain surfaces. However the texture is an uncomfortable mixture of stone scree and heavy clay, and although a welcome break from the thin-soiled mountains for the ancient farmer, they still represent second-class land. Hence my Argolid observations as to the rarity of activity traces from archaeological survey on this formation. In our Boeotia survey of summer 1979 we were able to demonstrate exactly the same situation.

The Younger Fill (Figure 2) is more confined in topography, generally comprising a riverside terrace above present streamcourses, except at the coast or in closed drainage basins, where it can expand to occupy very extensive areas, such as the great coastal Plain of Troy (Figure 3). It is exclusively alluvial, a well bedded silt with gravel lenses, brown to grey in hue, and usually only a few metres deep. The formation grades to a coastline very similar to the present one; it is a fresh, barely weathered deposit, and indeed associated artefacts, structures and C14 dates, demonstrate that it is a sediment primarily laid down from Late Roman times through into the early modern period. However it is not being deposited at present and contemporary streams are incising it, probably for the past two centuries or so. This formation is more difficult to account for, though most people are agreed that increased supply of bedload led to stream aggradation. Vita-Finzi and I have suggested that climatic change was the most significant factor, and certainly there is growing evidence from independent sources throughout Europe for an unparalleled decline in climate during much of this period (e.g. Alpine glacier advances). But there is also scope for human assistance, by exposing soil, via deforestation and over-cultivation, to the increased rainfall and runoff of this era. Nonetheless it is most striking that the last few centuries have seen the <u>end</u> of aggradation and an apparently lighter bedload promoting bed incision, despite undeniably the greatest human use and misuse of the Mediterranean landscape; and this coinciding with a return to a milder, less erosive climate.

The drab hue of the Younger Fill is considered to be due to reducing conditions as winter rain remained stagnant along the water-courses and under a more flourishing valley vegetation than today. Now the arrival of this formation was just as significant to Mediterranean cultures as the Older Fill, for although less extensive except at the coast, its light silts are ideal for cultivation, and being right beside present streamcourses are very amenable to irrigation. Indeed the fill has become the vital zone of commercial cash-cropping throughout the Mediterranean. Yet there is little doubt from careful study of geomorphic conditions when the Younger Fill was being deposited, that this fine alluvium could rarely be used to its excellent potential until recent centuries. The aggrading rivers sweeping all over the plains, and the presence of stagnant ponds, meant the loss of that land to farming and an ideal breeding ground for malaria until natural incision and human technology reclaimed the fill.

Fig. 3

Begging the question of why so much extra bedload became available for the two fills, we can securely state that the consequent stream reaction was the natural one, to adjust grade, i.e. long profile angle to the sea, so preserving equilibrium between bedload and available stream energy for transporting that load. By dumping sediment the streams raised the bed and steepened the long profile, thus bringing gravity to increase stream energy and also disposing of some of the extra load in the aggraded deposits. Between fills and since the last one, bedload has decreased and streams adjusted to a flatter profile by incising their bed through earlier fill. Today a typical Mediterranean streambed contains little but pebbles and coarse sand as far as its estuary, a strong contrast to the silts and clays it deposited throughout its catchment under earlier regimes. Clearly the absence of pre-modern sites on Younger Fill in archaeological surveys should now come as no surprise.

The Vita-Finzi scheme has not been without its critics, though much of the criticism seems to rest on misunderstanding of his geomorphic theories. Davidson in 1971, as geomorphologist to Renfrew's Sitagroi excavation in the Drama Plain, recorded active fan development there in prehistoric times, but by this he meant stream incision as predicted by Vita-Finzi for this period (Davidson, 1971).[5] More comprehensible was the error of Eisma, who already in 1964 found fault with Vita-Finzi from his studies of the great river valleys of western Turkey (Eisma, 1964).[6] Here we are fortunate enough to have plentiful historic and archaeological data for coastal plain development in the last 3000 years or so, especially linked in to major ancient cities. Eisma's point is very obvious from the following illustrations (Figure 4): the plains are visibly expanding at the expense of the Aegean Sea at a phenomenal rate (and their length is in tens of kilometres). Furthermore it is clear that the date of maximum expansion is Classical and Hellenistic rather than Medieval, more suitable for the contemporary massive population and intensive land use producing erosion, in contrast to the Dark Ages of the Vita-Finzi dating when those factors were at a minimum and climate far more plausible a factor.

Yet once again we are looking at a situation that is perfectly in accordance with the geomorphology of Vita-Finzi, as may be seen in this simple model (Figure 5) for an idealised Mediterranean valley. Stage One shows deposition of the Würm age Older Fill, and sea level well below present. Stage Two shows the early to mid Holocene with sea rising to near present, less bedload so streams incising into the Older Fill. This is referred to here as the Deltaic Phase, as it is typified by river deposition virtually confined to coastal delta situations; this lasts until the onset of the Younger Fill in Late Roman times (Stage Three), and aggradation then proceeds till c. 200 years ago when renewed incision and delta buildup occurs (Stage Four). The two phases of non-aggradation but delta build up are particularly interesting: if rivers are incising, what happens to the eroded material, and also to the reduced bedload still being fed to streams? Clearly the stream is acting to reduce its grade and is very efficient, so very little is to be deposited before the estuary is reached, and here indeed the vast mass of bedload will now be dumped. Yet to achieve the desired low angle stream profile, this dumping must not hinder continued stream access to fresh dumping grounds, so the desired nature of the estuary is a very low delta, which is added to by <u>lateral</u> expansion rather than vertical accretion.

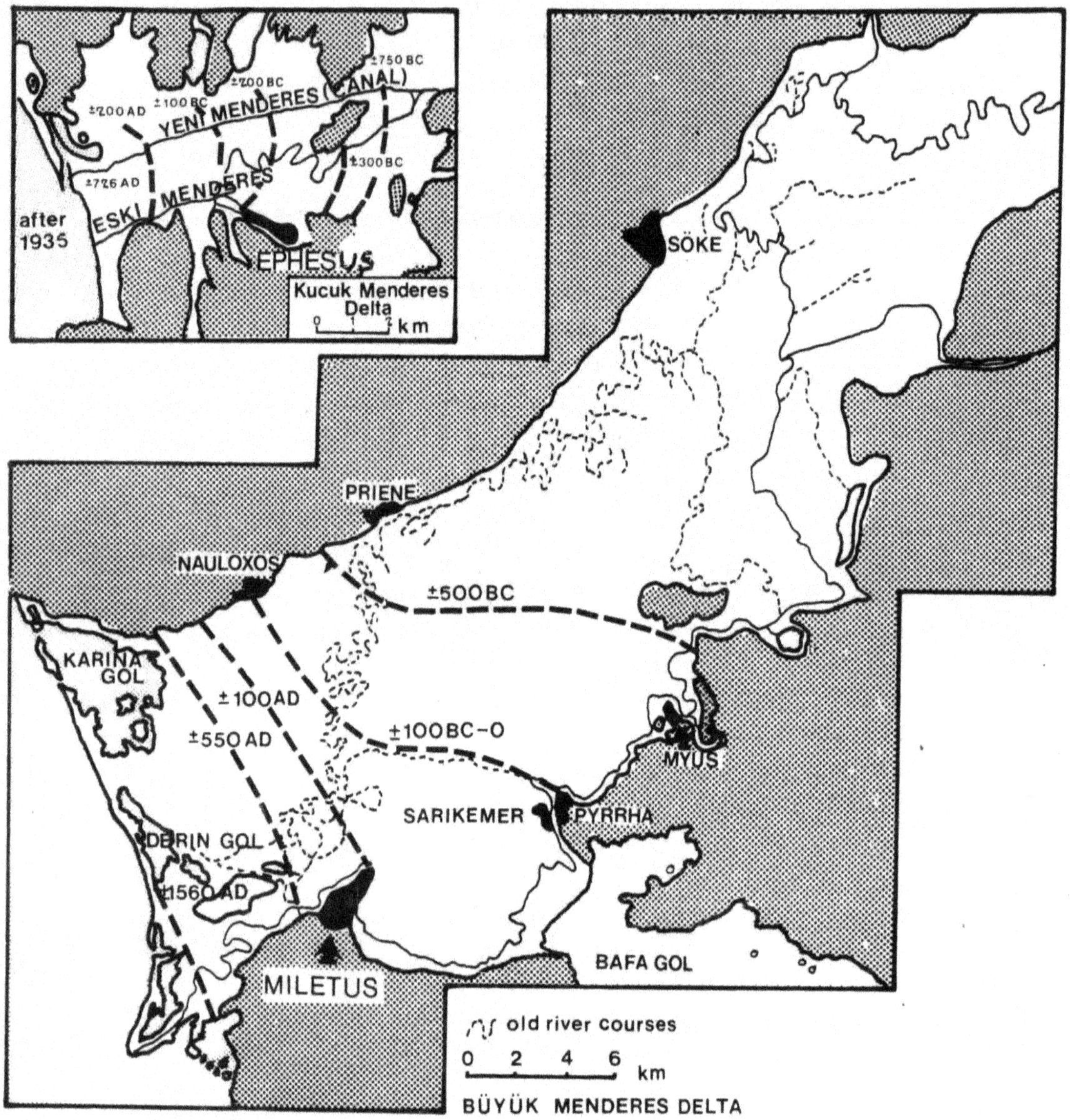

Fig. 4 (after Eisma, 1978).

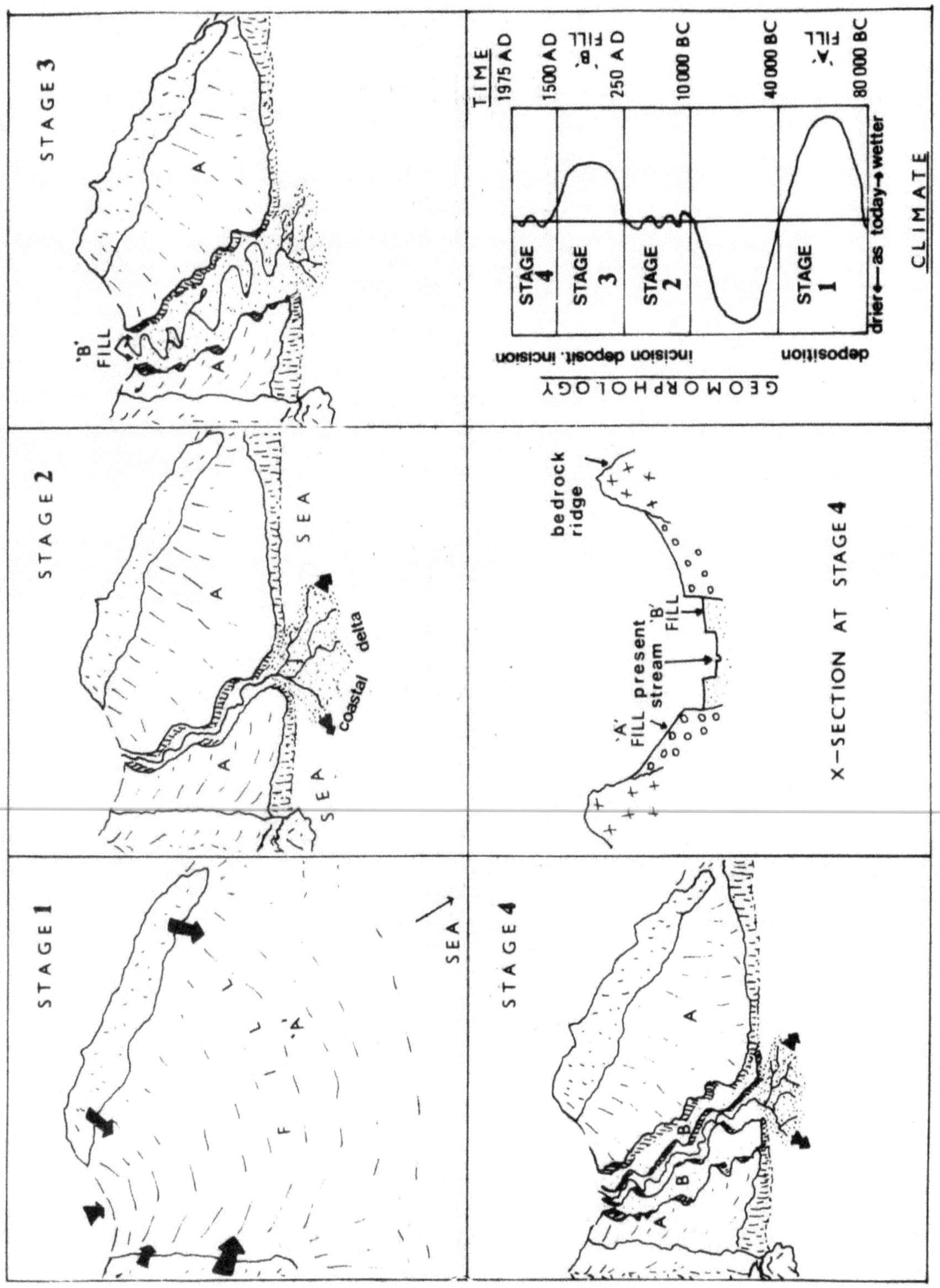
STAGE 1
'A' FILL
SEA
STAGE 2
A
SEA
coastal delta
STAGE 3
'B' FILL
A
STAGE 4
A
B
X-SECTION AT STAGE 4
bedrock ridge
'A' FILL
present stream
'B' FILL
TIME
1975 AD
1500 AD
250 AD
'B' FILL
10 000 BC
40 000 BC
80 000 BC
'A' FILL
GEOMORPHOLOGY
incision
deposit.
incision
deposition
STAGE 4
STAGE 3
STAGE 2
STAGE 1
drier←—as today—→wetter
CLIMATE

Fig. 5

So the expansion of Mediterranean deltas in the Vita-Finzi scheme should in fact be out of phase with the periods of river aggradation, of fill deposition, and the very slight delta growth after Late Roman times in Eisma's plots is wholly in accordance with prediction, just as the notable pre-Roman expansion fits too. The corollary of this, is that whilst in pre-Roman times these coastal plain cities would have had problems with harbour silting, though their hinterland valley fields still flourished free from floods and silt, after Late Roman times the deltas would have slowed or ceased growth but the whole plain behind the coast would have been swamped and virtually lost to cultivation by rapid valley alluviation upwards. We do in fact know this precise sequence from these very plains: Ephesus city, for example, was completely buried by alluvium, up to 8 metres deep in places, precisely dated from Late Roman to early modern times. In fact Eisma has very recently published a paper acknowledging this major critical error regarding the Deltaic phase.[7]

Such massive landscape changes in flourising ancient city territories cannot have been without cultural and economic repercussions. Foss, in a recent paper on the collapse of urban life in Turkey in the later Ist millennium A.D., finds foreign raids and piracy insufficient reason for a disintegration of Byzantine civilization almost as dramatic as in the conquered western Roman empire. He toys with other factors such as climatic change.[8] I think what we are looking at in these valleys must now be considered a major element in this cultural collapse.

However, one further problem may be apparent from this illustration, and this can be tackled with a further example, a study of the plain of West Macedon, just by Thessaloniki (Bintliff, 1976).[9] (Figure 6). The interest here was in the Holocene development of this vast alluvial plain, and with the aid of field studies, some laboratory sample analysis, excavated sections from the Neolithic site of Nea Nikomedeia, and plentiful historic descriptions, the following developmental sequence for the Plain landscape was proposed: Stage A, a great ocean gulf intrudes upon a highly fertile, well-drained plain of recent marine and lacustrine silts, upon which at some distance from the sea has been established an early Neolithic farming village at Nikomedeia. Stage B covers the period from c. 500 B.C. to 0 B.C., and shows the rapid progress of delta extension into this sea inlet. By Stage C, Late Roman times, the inlet is reduced to a poorly-drained lake, and the city of Pella, in 500 B.C. on the sea, is now approaching the present day figure of c. 30 kilometres from it. By Stage D, c. 1900 A.D. the Younger Fill has been deposited and the whole hinterland of the plain is also alluviated, while at the same time the low delta lands raised up high by aggradation. The Plain today is a recently-drained, highly fertile alluvial formation of vastly different character even to its ancient form. Now note here again, this rapid, 1st millennium B.C. speeding up of the delta advance, as in western Turkey, —its virtual cessation in Medieval Younger Fill times, then recent readvance. Why did not deltas push out at a uniform rate through the Holocene?

The answer throws us metaphorically and literally into the deep end of Quaternary studies: sea level changes. Since no-one can offer a generally agreed universal eustatic curve, I suggest that the best we can do is to construct large regional syntheses as working blocks for some future eustatic architect to fit together. Having worked through most of the relevant sites

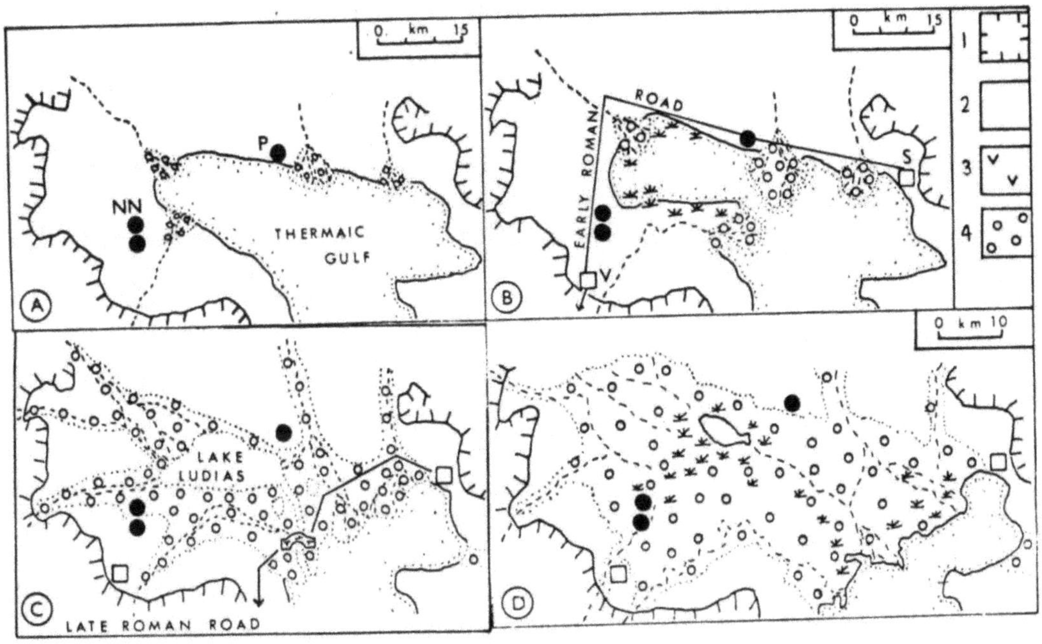

Fig. 6

Fig. 7 Submerged Late Roman site at Trinisia, southern Peloponnese.

in these Greek study regions (Figure 7), and through most of the relevant publications on Mediterranean sea level change, an unexpected result emerged: that the older authorities were right in arguing that around the Mediterranean the general tendency is of continuing coastal submergence from prehistoric times, at the rate of c. 1 metre per millennium. Even more surprising, the largest body of evidence, that studied by Flemming, confirms this conclusion despite his totally contrary interpretation. This conclusion, however, is a rarely favoured option on a cross-section of proposed sea level change curves (Figure 8) which usually have present sea level reached c. 4-2000 B.C., then little change, or slight ups and downs since then. This 1 metre rise figure is offered here as a regional generalisation, and whether this is not truly an eustatic phenomenon but due to tectonic downwarp or other factors, is as yet uncertain. But it does give us a working figure. Rather greater agreement is found with the shape of the eustatic curve, with a rapid ocean recovery after c. 17000 B.C. (around 10 metres rise per millennium), then a pronounced slowing down after c. 7000 B.C., to 1 metre per millennium or so till present or not long ago. It is this shift from fast rise to slow rise that is the main key to solving the problem of Mediterranean coastal delta development. For it seems fairly clear that the sea was rising far too rapidly in the early Holocene for delta infill to keep pace, so deposition was unable to create large subaerial deltas; once the slow rise had taken over, however, from Neolithic times, subaerial delta sediment could accumulate, but there still existed in most cases a deep offshore basin to fill, since the coast generally sloped steeply down to the minus 100 metre low Würm sea level (as can be seen from plots of the angle of Older Fill disappearing underwater today). So it would usually be a significant period into the slow rise era before this offshore basin was filled and delta sediment could push out into the sea above the water. In the cases studied, it is clear that notable delta expansion is a feature only of the Ist millennium B.C.; during the Ist millennium A.D. and a good part of the 2nd, of course, delta extension was limited since hinterland aggradation prevailed, and only recently has it resumed. That sea rise is continuing is often demonstrated by the truncation of the Younger Fill formation at the retreating coast today, where it has been starved of deltaic sediment supply.

Given the complex factors that can influence the geomorphic history of small drainage systems, such as landslips, human interference, tectonic changes etc., and even variations in climatic history between sub-regions, it would be surprising if all Mediterranean valleys showed the complete Vita-Finzo sequence with precisely the dates he published from his Mediterranean-wide researches (Figure 9); however the present work in Greece has revealed very few exceptions to his scheme and chronology, and this includes my ongoing Boeotia Project. A survey of the published literature provides widespread confirmation of this work. However, a number of papers[10] contain additional or alternative dates for recent alluviational episodes around the Mediterranean, though in almost all cases the archaeology is highly questionable. One of the few papers that provides a clear case study, well published, disputing the Vita-Finzi chronology, is the recent one of Davidson and Tasker on alluviation on the Greek island of Melos (Davidson et al., 1976).[11] They recognise a recent alluvium of Younger Fill type, but suggest an earlier date of inception, favouring a man inspired erosional episode. My own research fieldwork on Melos produced about a score of Younger Fill exposures with Roman and post-Roman

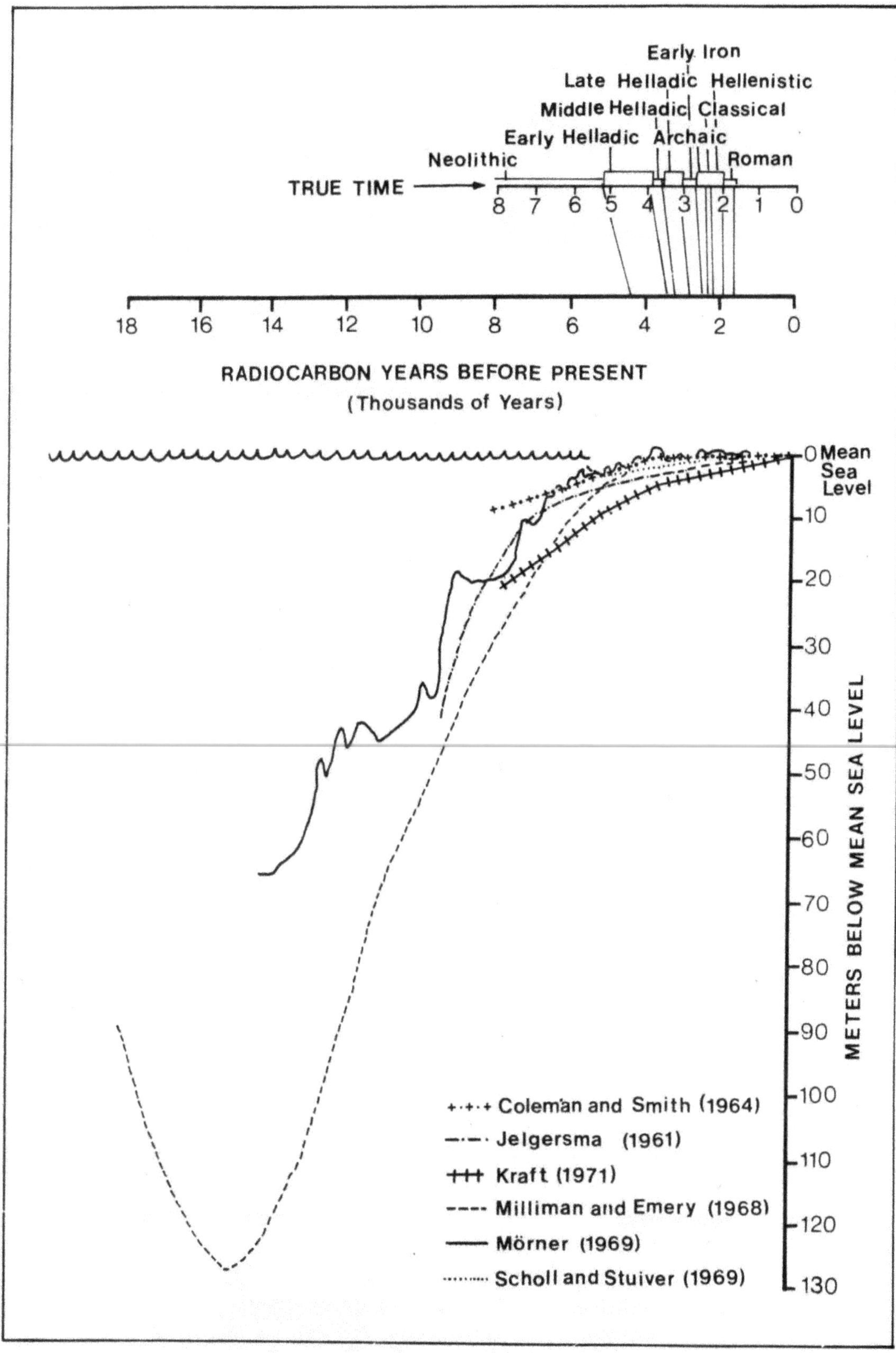

Fig. 8 (after Kraft).

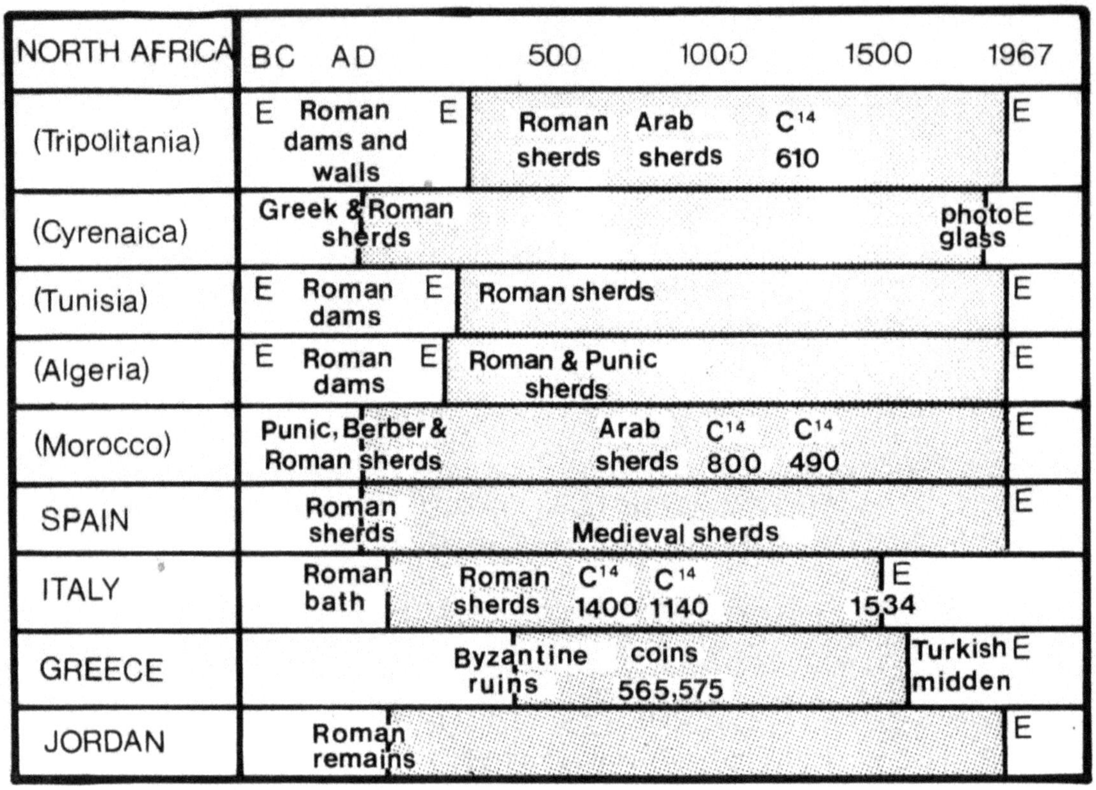

Fig. 9 The Vita-Finzi chronology for recent Mediterranean alluviation (after Vita-Finzi 1969).

Fig. 10

Fig. 11

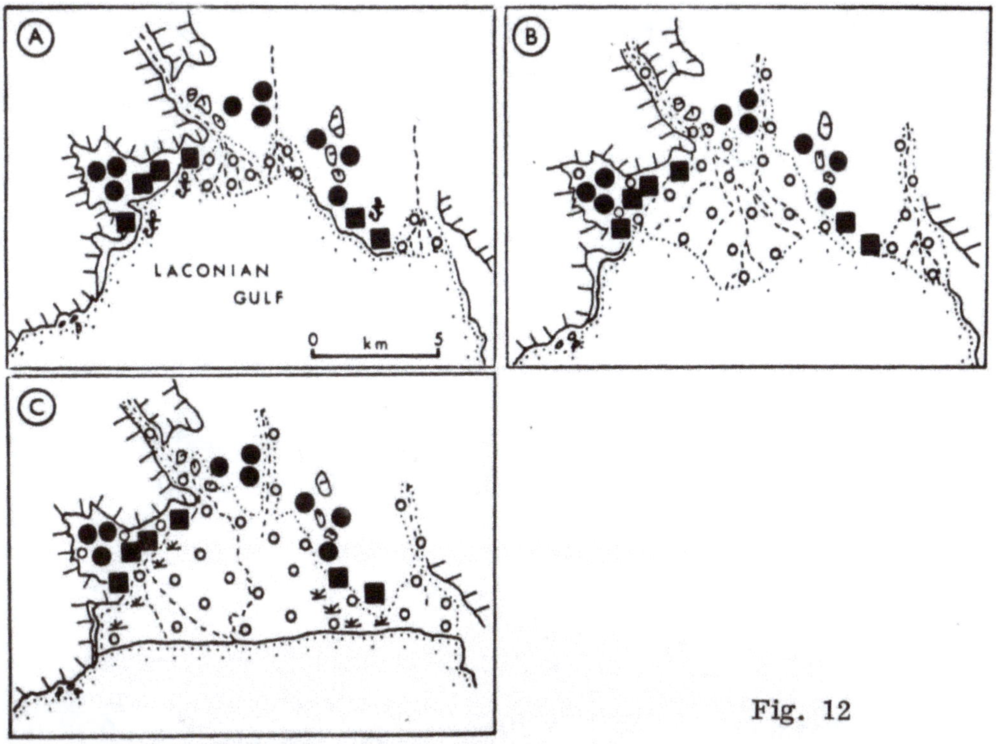

Fig. 12

dating associations, and having seen the two exposures discussed in their paper, I do not feel that the new evidence requires any re-think of the Vita-Finzi chronology: the Phylakopi coastal plain fill, I suggest, shows basal beach deposits, then a level of Bronze Age hillwash from the Phylakopi site above the test trench, covered by extensive Younger Fill alluvium of Roman and later date; the published upstream section may be seen as two deposits of colluvial hillwash separated by an organic level with C14 around 1000 B.C. —not really Younger Fill Material at all.

The complexity of recent landscape change for interpreting archaeological site environments is well illustrated by the Helos Plain in Mainland Greece (Figure 10), at present a flat alluvial expanse. A large number of ancient and prehistoric sites have been found, but their location usually has the following topography (Figure 11): an island, often tiny, of well-drained Tertiary marine silts above still inadequately-drained alluvium, Careful field study suggests the following sequence (Figure 12): a prehistoric, well-drained fertile low hilland around an ocean inlet; by Roman times the delta extension is well advanced; finally, in Medieval times we see the complete swamping of the whole district by Younger Fill aggradation, till only the tops of the hilland plus sites peep up above the alluvium.

A good illustration of the economic effects of these changes comes from the Argos Plain, and also Mainland Greece (Figure 13). The prehistoric and ancient settlements marked, are concentrated on first class quality arable hilland with the occasional coastal station, and no other sites on the second class Older Fill of the main plain. Also mapped is the expansion of the delta up till the Classical era, then on the adjacent map the transformation of the lower plain as it is built up by Medieval Younger Fill aggradation. A notable phenomenon is the Bronze Age city of Tiryns, once on the shore, now two kilometres away, with the lower town buried by many metres of alluvium (Figure 14). Note also (Figure 15) the recolonisation in the recent period of the new alluvium, into what becomes the new irrigated agricultural heartland. By the early 20th century of this illustration, most of the population (with size of settlements represented by circle diameter) dwell in or by this new land, and the ancient foundations on the plain edge in the hills are becoming mere backwaters.

Tantalising references in classical geographers led me to visit the region of ancient Troy in 1977 (Figure 16). The region as a whole is largely composed of old dense rocks left to forestry and grazing, and arable food production is concentrated on two linked basins, the coastal plain of Troy and the inland Ezine basin. Fundamentally they consist of fertile marine silt hilland and recent alluvial bottomland. As predictable, the present incising river regime cannot be seen as responsible for the alluvial plain formation, which is of typical Younger Fill aggradation character. Furthermore, ancient references and Homer's description in the Iliad call for a deep harbour inlet at the coast, where now an exposed shore is found. I was able to use detailed field observations, data from shallow borings and earlier studies (that noted features such as recent sea cliffs in the plain hinterland), together with fairly full historical references and limited archaeological data. Ancient sources included distances from Troy to the sea at varying points in time, to which one could add estimates over the last 150 years for the rate of expansion of

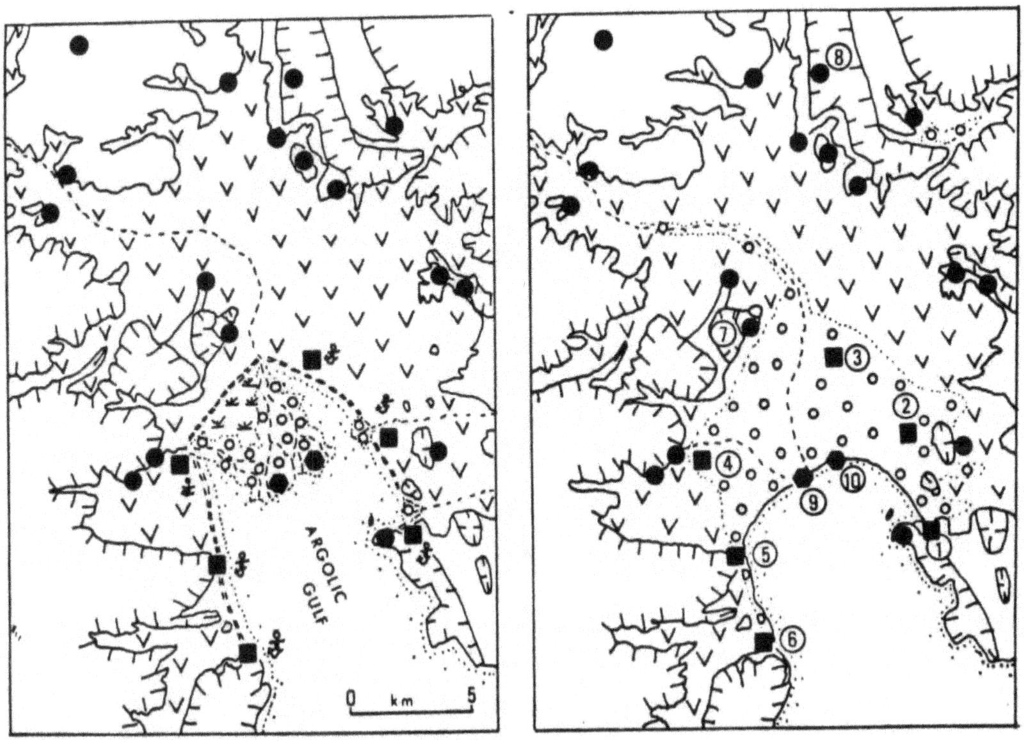

Fig. 13

Fig. 14

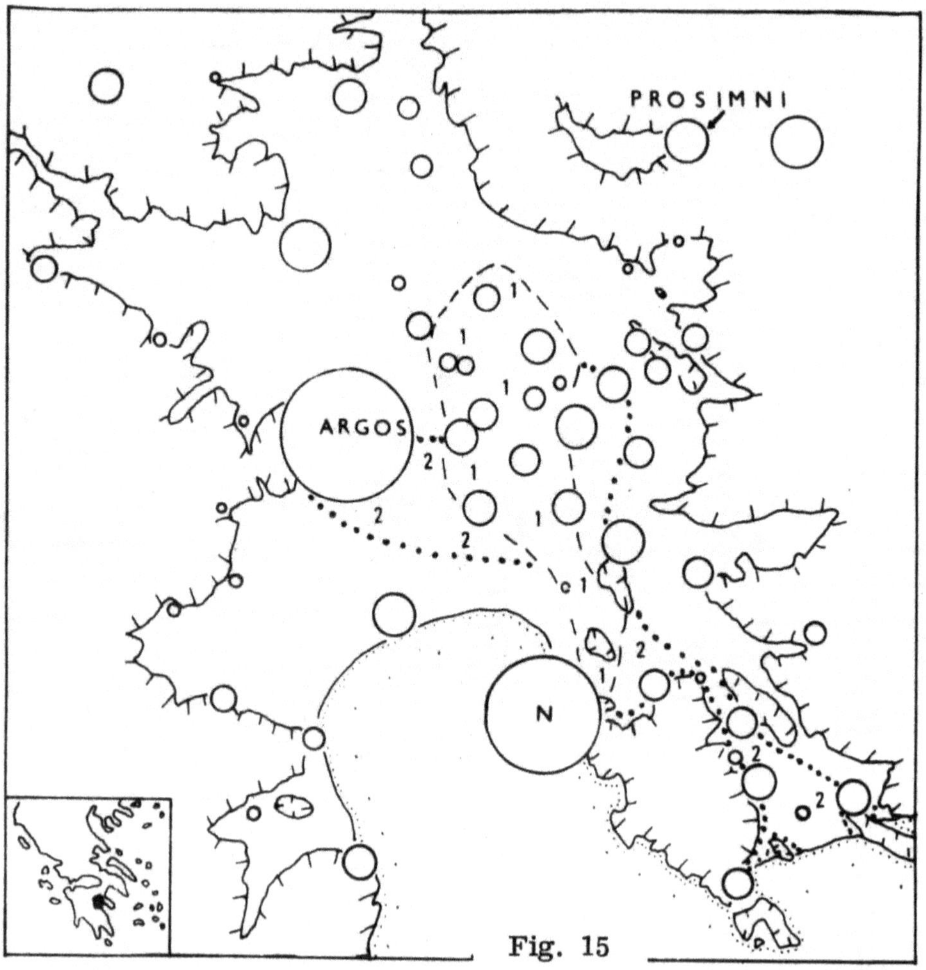

Fig. 15

Fig. 16 Block diagram of the Trojan Plain, the Ezine Basin and their upland hinterland (after Bilgin 1969).

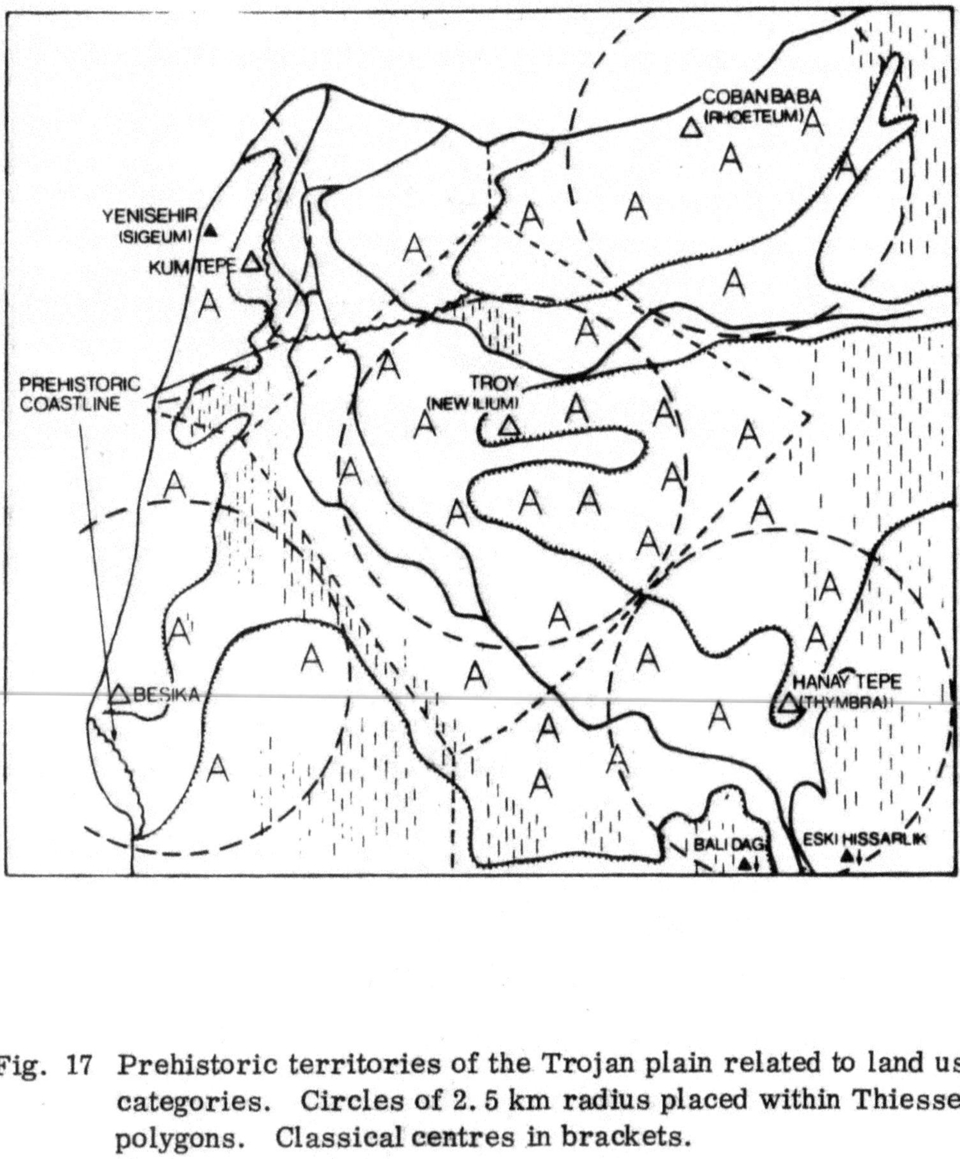

Fig. 17 Prehistoric territories of the Trojan plain related to land use categories. Circles of 2.5 km radius placed within Thiessen polygons. Classical centres in brackets.

▲ Minor prehistoric sites
△ Major

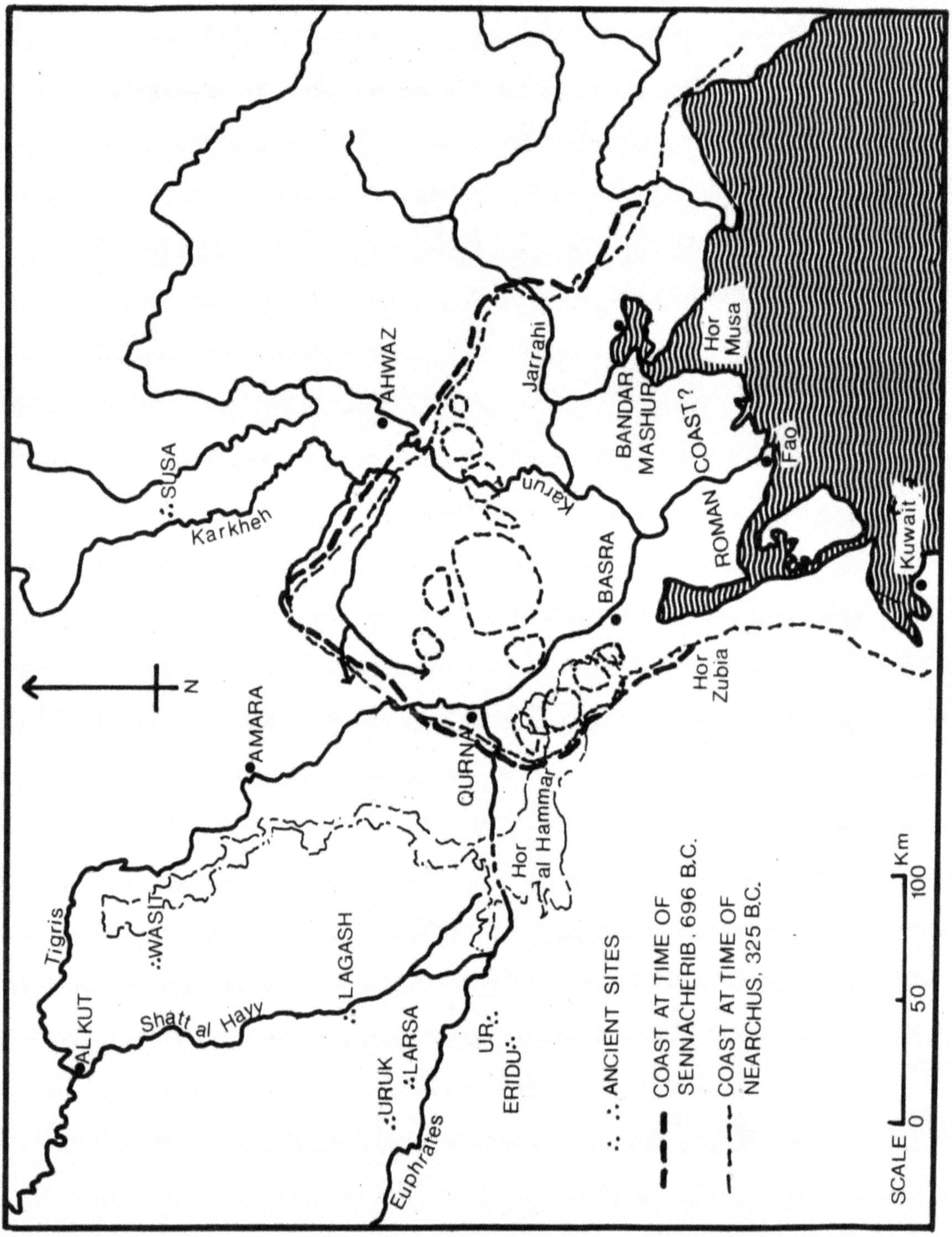

Fig. 18 Past shorelines of the Mesopotamian Delta, based on De Morgan (1900) and recent publications.

the delta. From all this it was possible to set up provisional period by period reconstructions for the development of the plain, from the age of the supposed Trojan War up till modern times. From this basis one could then begin to study the human ecology of individual periods. In Figure 17 we can see the major prehistoric settlements and their classical successors forming a recurrent network of land exploitation, with Thiessen polygons appearing to isolate naturally viable territories of good land (marked A) amid poorer land (dashed). The ancient coast is approximately indicated. The size and richness of the individual Thiessen cells compares well to the estimated importance of the settlement within it and to modern village densities: Troy understandably is best off for land, with four modern villages within its cell. An additional result was to be able to suggest that Homer's descriptions of the plain were close enough to embody an eye-witness account of the pre-Classical landscape.

Moving on briefly to other people's work, at long last it seems that the history of the Mesopotamian delta is being unravelled (Figure 18). De Morgan argued in 1900 that 7th and 4th century B.C. texts implied a subsequent massive delta expansion into the Persian Gulf; archaeologists were soon to point to Mesopotamian Bronze Age texts associating cities such as Ur and Eridu with the coast, now hundreds of kilometres distant. All this was supposedly quashed in a definitive study from Lees and Falcon in 1952,[12] who put a strong case for a long-term balance between deposition of sediment and tectonic subsidence in the delta region, producing minimal change in the position of the coastline since Bronze Age times. Much subsequent discussion centred on unpublished 1930s corings through the plain by Macfadyen, which were finally published last year. Several articles of 1978, using these data and older published material with similar implications, present a convincing case for something like the old De Morgan view,[13] which should not surprise us in the light of the remainder of this paper. Firstly, detailed tectonic studies do not support significant subsidence of the delta area during the Holocene (and indeed the C14 date of 8000 b.c. for the transgressing Gulf waters at c. 32 metres below present sea level is well within the range of eustatic figures from around the world). Most of the Gulf was indeed dry in glacial times, floored by the Dibdibba Formation, Miocene to Pleistocene in age and typically steep in grade at its surface, planed to low Würm sea levels. Above this formation the corings reveal the transgressing marine and estuarine Hammar Formation, which shows that the early Holocene Gulf formerly extended up to 250 kilometres inalnd of the present coastline. The upper facies of this formation show, however, the gradual retreat of purely marine sediments and a replacement by estuarine and ultimately freshwater alluvial deposits; here we can surely see the gradual conquest of the submerged delta area by alluvium with the slower eustatic regime, and the infill of the basin floor, then the rapid push forward in historical times of the subaerial delta proper. From a 3rd milennium coast near Ur, we move to the 7th and 4th century B.C. references that still suggest considerable openwater not far from Ur, then by Roman times the delta, in places, seems to have been close to its present extent. The full implications for the development of Mesopotamian civilisation should prove a fascinating field for research.

A number of recent papers on the Po delta seem to suggest a similar sequence:[14] firstly, the delta advance over the last few centuries is far too rapid for the overall estimates of reclaimed land since Etruscan times, and would thus be appropriate for a reduction in delta accretion for Younger Fill times. Secondly, we have a date of c. 6000 b.c. for the transgression of the Adriatic into the area of the present Po plain, while marine influence seems dominant in delta sediments till c. 1000 b.c. when freshwater deposition takes over. The general matching of this late date meshes very well with our chronology for subaerial delta takeoff elsewhere.

Several conclusions from this line of environmental archaeology are offered as follows:

(1) If Catchment analysis is allied to Quaternary Science on a regional basis it is a highly informative approach to Man-landscape interactions in the past, and should be embraced by all environmental archaeologists.

(2) Holocene landscape change suffers from wide neglect, which is why environmental archaeologists still have to go out and fill in on the post-Tertiary geomorphology. Moreover, local sediments can often be fruitfully linked in to more general models of stream and slope behaviour on the scale of a complete river catchment or wider region. Not only is archaeology generally essential, with history, for obtaining a chronology for such changes, but it is becoming increasingly clear that the dynamics of landscape change were of vital importance to land use and cultural development and had serious repercussions on past societies.

(3) Although we are rightly aware, in our Ecology Crisis age, of the negative side of human activity in the landscape, we must not assume that nature is passive—the possibility exists of far-reaching earth processes in which human influence is not a crucial factor.

(4) Recent landscape change is complex, both spatially and temporally: the heavy erosion that produced the Younger Fill removed much cultivated Older Fill and useful hill soil, and the period of deposition was one of great problems for the farmer. Yet the new area of alluvial deposition, due both to nature and to man, was a net gain for Mediterranean agriculture in the longer term. Our denuded English Chalklands, the focus of recent interest in the Bronze Age shift from them to the true lowlands, provided likewise eroded material to enrich the Chalk vales and the major river valleys further downslope, into which the diaplaced settlers were moving.

NOTES

1. E. S. Higgs, and C. Vita-Finzi, 1970: 'The Prehistoric Economy of the Mt. Carmel Area', Proc. Prehist. Soc. 36: 1-37.

2. M. R. Jarman, 1972: 'A territorial model for archaeology', pp. 705-739. In Models in Archaeology, edited by D. L. Clarke.
   G. Barker, 1975: 'Early Neolithic land use in Yugoslavia', Proc. Prehist. Soc. 41: 85-104.

3. D. Webley, 1972: 'Soils and site location in prehistoric Palestone' pp. 169-180. In Papers in Economic Prehistory, edited by E. S. Higgs.

4. J. L. Bintliff, 1977: Natural Environment and Human Settlement in Prehistoric Greece. British Archaeological Reports, Suppl. Series 28; and 'New approaches to human geography: Prehistoric Greece, a case study', pp. 59-114., in The Historical Geography of the Balkans, edited by F. W. Carter, 1977.

5. D. Davidson, 1971: 'Geomorphology and prehistoric settlement of the Plain of Drama', Revue de Géomorphologie Dynamique, 20: 22-26.

6. D. Eisma, 1964: 'Stream deposition in the Mediterranean area in historical times', Nature, London, 23: 1061.

7. D. Eisma: 'Stream deposition and erosion by the eastern shore of the Aegean', pp. 67-81. In The Environmental History of the Near and Middle East, edited by W. C. Brice.

8. C. Foss, 1977. Archaeology and the twenty cities of Byzantine Asia. AJA 81: 469-486.

9. J. L. Bintliff, 1976: 'The Plain of Macedon and the Neolithic site of Nea Nikomedeia', Proc. Prehist. Soc. 42: 241-262.

10. e.g. J. Torrent and C. Roquero, 1974: 'The occurrence of an argillic horizon", Trans. of the 10th Int. Conf. of Soil Science 6(2): 354-361; or R. A. van Zuidam, 1975: 'Geomorphology and archaeology: evidence of interrelation at historical sites in the Zaragoza region, Spain', Zeits. f. Geomorphologie 19: 319-328.

11. D. Davidson et al. 1976: 'Erosion and prehistory in Melos', J. Arch. Sci. 3: 219-227.

12. J. de Morgan, 1900: Délégation en Perse, Mém. 1, Paris; G. M. Lees and N. L. Falcon 1952: 'The geographical history of the Mesopotamian Plains', Geog. J. 118: 24-39.

13. e.g. W. A. Macfadyen and C. Vita-Finzi 1978: 'Mesopotamia: The Tigris-Euphrates delta and its Holocene Hammar fauna', Geol. Mag. 115(4): 287-300; or J. F. Hansman, 1978: 'The Mesopotamian delta in the 1st millennium, B.C.', Geography, July, pp. 49-61; or Larsen and Evans, 1978: 'The Holocene geomorphological history of the Tigris Euphrates-Karun delta', pp. 227-244, in W. C. Brice, (ed.), The Environmental History of the Near and Middle East.

14. J. Bethemont, 1974: 'Des travaux et des hommes dans le delta du Po', Revue de Geographie de Lyon, 49: 253-284; G. C. Bortolami et al. 1977: 'Land, sea and climate in the northern Adriatic region during late Pleistocene and Holocene', Palaeog.,Palaeoecol., 21: 139-156.

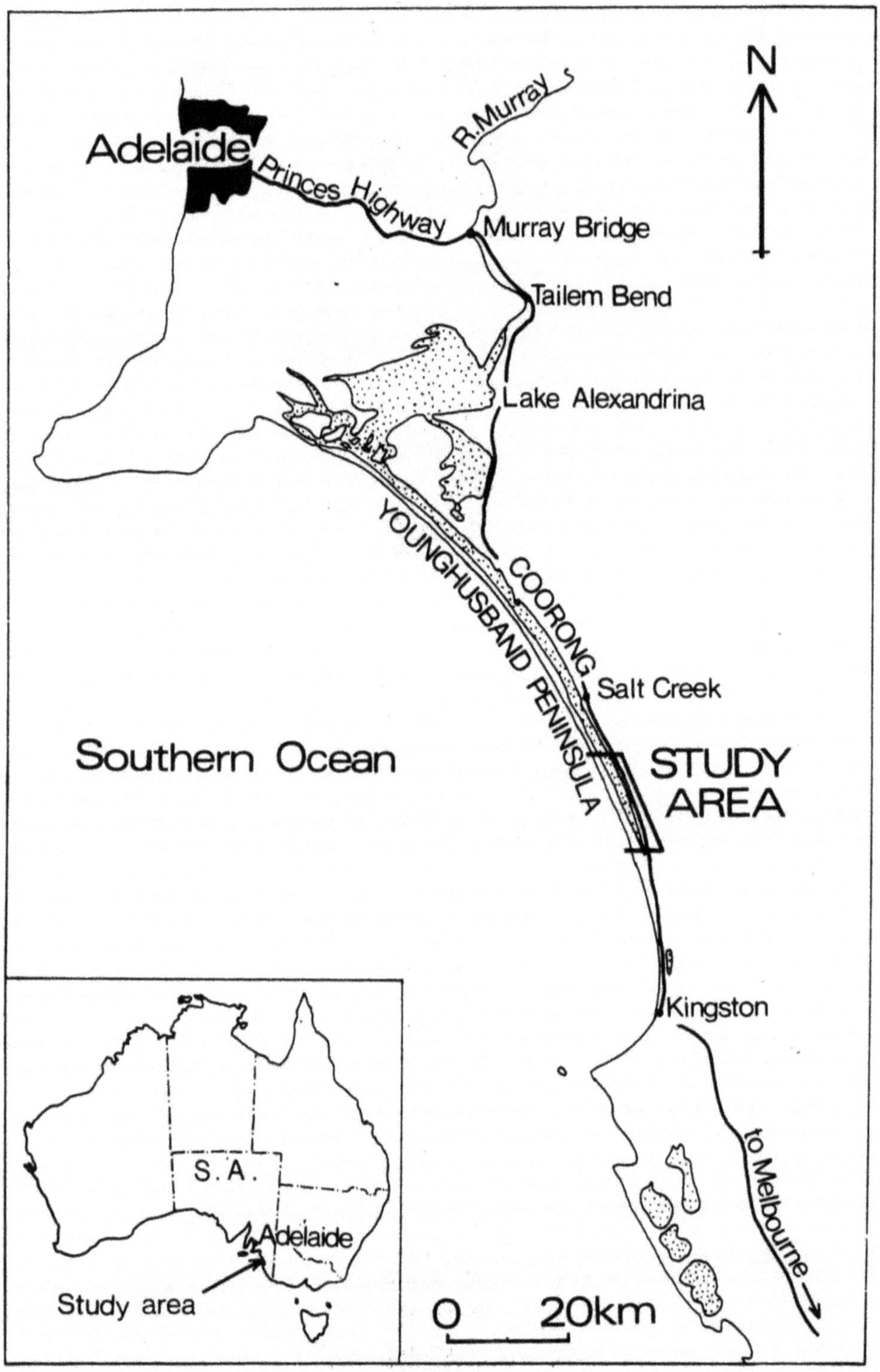

Figure 1. The location of the Younghusband Peninsula, South Australia.

THE INTERPRETATION OF STABILITY AND INSTABILITY IN A MAJOR COASTAL DUNE SYSTEM

D. D. Gilbertson
Department of Prehistory and Archaeology
University of Sheffield

1. INTRODUCTION

The nature of the relationships between human activity and earth surface processes acting in historic and pre-historic times may be of immediate, even crucial significance for the contemporary management of important natural and cultural resources. Similarly satisfactory appreciation of the ecological and geomorphic significance of the earlier occupants of an area can only be gained when armed with adequate data on the sensitivity of that land resource to contemporary pressures.

The Lower Younghusband Peninsula of South Australia (fig. 1) is a major coastal barrier system, now characterised by very active transgressive dune systems. It is important to determine the extent to which current accelerated dune erosion and drift is a legacy of past human activity and environmental change as opposed to being the result of contemporary land usage. These interprative problems are shared by many other major dune systems along the Australian coast, and are described by Bird (1973) and Thom (1974).

Active sand sheets may represent up to 95% of the surface area of the Peninsula. The proportions are somewhat lower in the main study area, where the mobile, or now stable vegetated parabolic dunes are separated from areas further inland by a discontinuous chain of lakes and alluvial lowlands known as the Lower Coorong or the Kingston Lake Chain.

Late Holocene sea level appears to have stabilised relative to the Coorong area at about 6,500 years B.P. (Von der Borch 1976) and the developing barrier gradually cut out marine access to the Coorong area, such that the only modern breach is at Murray Mouth where the River Murray and the Coorong reach the sea via Murray Mouth (fig. 1). The coastal dunes are almost entirely of carbonate sand derived from the remains of marine organisms.

The region has a semi-arid climate with marked summer drought, which potentially places some sand binding herbaceous and shurb species under stress. Previous extensive usage of the dunes by the aboriginal Tangannekald people is described by Tindale (1938), and attested by the extent of midden covered palaeosols, which can cover up to 40% of the total dune surface. The direct and indirect consequences of previous European pastoralism is indicated by a number of derelict homesteads, the massive populations of rabbits, feral cats and foxes, and the abundance of European and South African plant species. Contemporary usage comprises wildlife study, peaceful recreation and the use of Off-Road-Recreational-Vehicles. (ORV's). By late 1977 the area from

Murray Mouth to the southern end of the study region had become a protected area under the control of the South Australian National Parks and Wildlife Service. Most pastured stock had been removed during the previous 20 years, very few remaining in the last decade. Recreational vehicle use has continued; since 1977, illegally.

2. STABILITY AND INSTABILITY IN THE DUNE SYSTEM: POSSIBLE EXPLANATIONS

There is a substantial and often confused literature dealing with the causes of stability and instability in this and similar dune systems.

Causes of Stability

1. One local interpretation suggests difficulties of interpretation may have been caused by inadequate perception of cause and effect. Observations by ORV users made at 7 to 14 days intervals over a 2 year period of intensive use failed to note change; however, dune advance was noted upon return to the area after an enforced 2 year absence from the area. Vehicle induced soil compaction effects might by inducing stability (RVCC n.d.). Possible beneficial compaction effects on soil moisture levels and plant growth have been detected in small scale studies elsewhere (Liddle and Moore, 1974; Liddle and Grieg-Smith, 1975, a, b). However caused, any increase in soil moisture levels is likely to be beneficial to dune stability in much of the dune system.

2. There is a persistent, oft repeated belief that the dunes are stable; the dune topography tends to remain the same.

Causes of Instability

1. Natural Factors

Singly or in combination:- Cyclic alternation of stability and instability have been reported in western European parabolic dune systems (Landsberg, 1956; Ranwell, 1972). More locally Easton and Radok (1970) have reported the presence of significant local annual and semi-annual fluctuations in the level of the mean sea level surface, which might be suggested as exacerbating the drift problem by accelerating the erosion of the frontal dunes of the Peninsula.

Tectonic activity might be important; earthquakes were reported at Mt. Gambier in the mid-1950s; whilst Bourman (1974) has indicated land subsidence might be invoked to explain recent erosion at Middleton, near Murray Mouth. Climatic shifts manifesting themselves as increased frequencies and magnitudes of storms causing greater wave attack are frequently mentioned (Bird, 1973; Thom 1974 for discussion). Similarly any change in soil moisture status resulting from variations in precipitation/evaporation ratios in this semi-arid area are likely to have profound effects on the viability, abunddance and productivity of the sand binding dune vegetation.

2. Anthropogenic Factors

Anthropogenic influences on dune stability may be separated into three groups:

a) Contemporary recreation;

b) European pastoralism;

c) Pre-European aboriginal land use.

a) Contemporary recreation—over the last 20 years recreational use by off-road-vehicles has increased dramatically, such that during the June (winter) 3 day long weekend in 1975, it is estimated some 230 vehicles and 500 people were in the National Park, south of Salt Creek.

b) European pastoralism has involved many activities likely to cause or exacerbate dune instability problems. The Peninsula once supported four homestead/dairies, with abundant cattle and sheep. In the late nineteenth century Cantara (fig. 4) was an important horse-breeding station, supporting 1500 horses. However, the spread of wild rabbits and wild dogs from introduction elsewhere caused a reduction in the number of horses to about 350 at the turn of the century (Paton, 1977). The Peninsula even supported an Ostrich farm, with 500 birds maintained in 1912. Overgrazing, trampling, burrowing and scratching are all important potential causes of dune instability. Aboriginal legends of the local Tangare people can be interpreted as indicating that extensive physiographic changes have been caused by European pastoralism.

> "Our ('maidawanli') told us long, long ago to "beware of ants". White man must be the "ants" he spoke of, for they have eaten away all my people, my game, and even my sandhills". (Tindale, 1938, p. 20).

c) Pre-European aboriginal land use. Firing the bush was extensively practised to drive animals onto sand ridges entering the Coorong, whence animals were speared. The area is also sufficiently bleak in winter to have necessitated relatively extensive use of skins for clothing by local aborigines (Tindale, 1938); accidental escapes of fire are likely to have been frequent. The reduction or removal of surface vegetation by fire is likely to have exposed considerable areas to deflation.

3. STUDY METHODS

The following field and laboratory procedures were practicable for studying this relatively remote region:-

1) The rate and pattern of recent sand drift was determined using available aerial phtography—1947 to 1967 (black and white, 1:50,000), 1975 (true colour, 1:10,000).

2) Archaeological remains and described landscapes dating from the period immediately pre-dating the main introduction of European pastoral activity in the area were located and mapped. Coupled with literary and cartographic data, "baseline" conditions in specific areas immediately prior to the European impact were reconstructed. The critical data concern the route and remains of the Adelaide to Melbourne telegraph line, surveyed in 1855; and abandoned in 1929.

3) The apparently "stable" areas were identified in field mapping as outcrops or exposures of prehistoric (i.e. pre-European) palaeosols and middens, and a semi-quantitative assessment of the resistance of vegetated dunes, transgressive dune sheets, middens and palaeosols to erosion was obtained by using

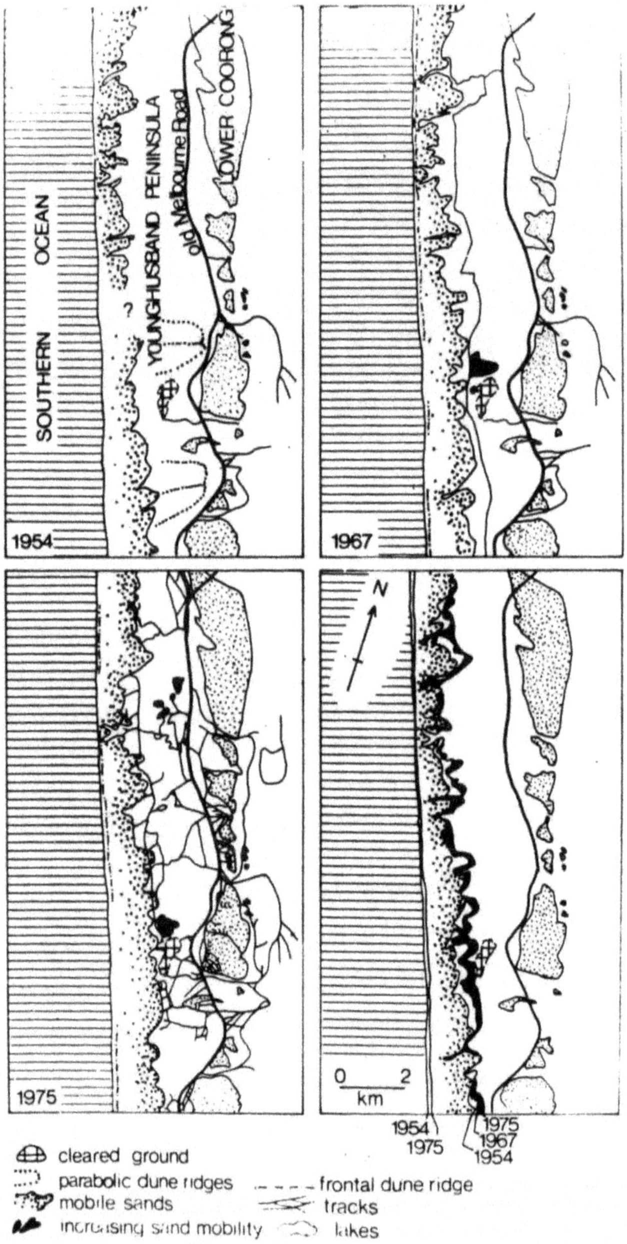

Fig. 2 Compilation of aerial photographs showing the position of the main mobile dune face and track density near Coolatoo, Younghusband Peninsula, 1954, 1967, 1975: (early coach and/or bullock track omitted, except where widened by post-1954 use). Redrawn and updated from Gilbertson (1977b).

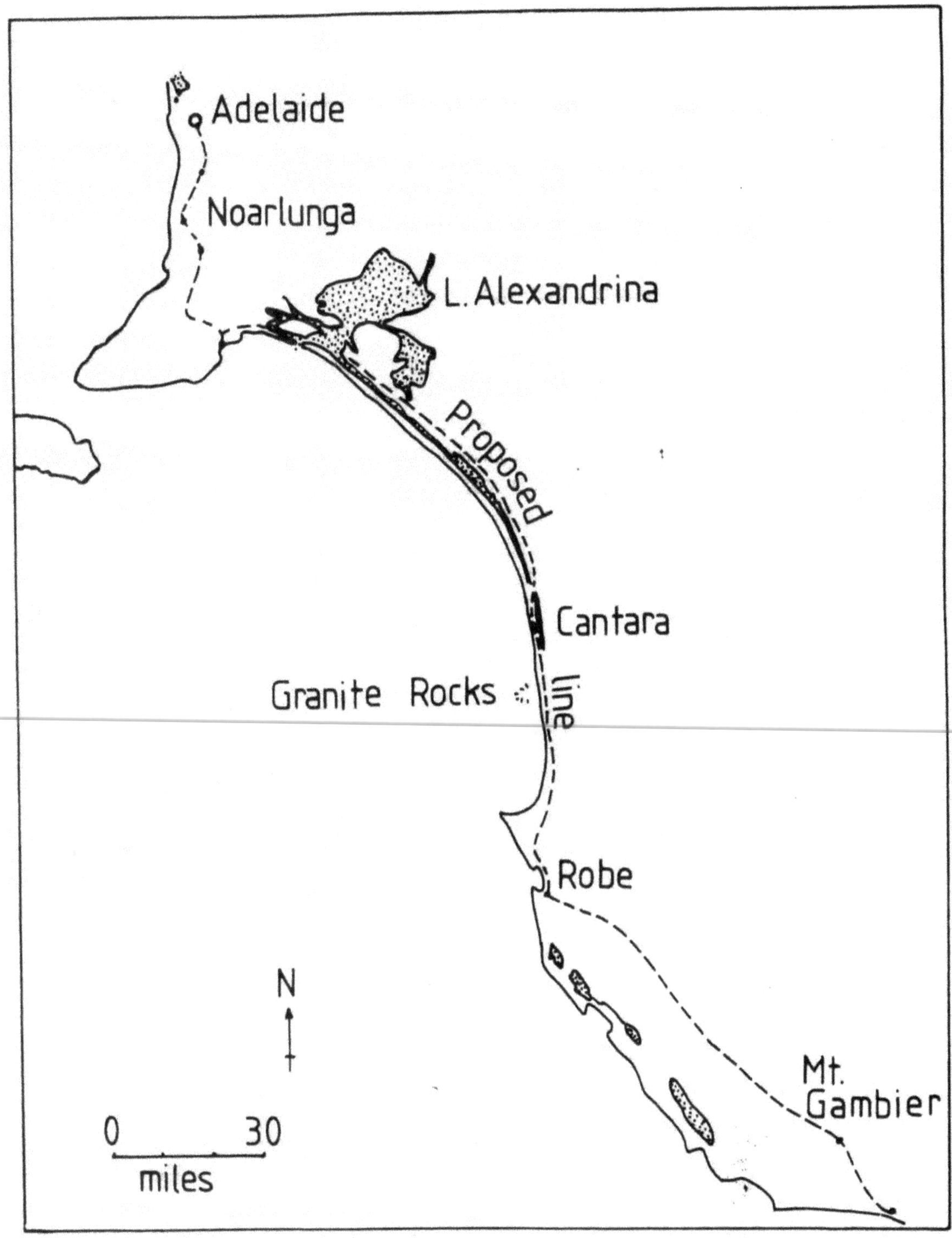

Figure 3. The route of the proposed telegraph line between Adelaide and Melbourne, surveyed by Charles Todd in 1855. Redrawn from Todd (1856).

a portable shear strength vane developed by the Department of Civil Engineering, University of Adelaide. The four bladed vane measured 0.152 m in height; 0.076 m in width. Whilst giving reliable shear strength results for cohesive soils such as clays, determinations are more difficult for non-cohesive sandy soils. Here, shear strength measurements are more affected by differences in confining pressure caused by possible prior conpaction, as well as changes brought about during the shearing motion of the vane during the process of measurement (Arnold, in litt). In the latter case the readings give a more qualitative picture of a soil's resistance to shear. Deflation by wind is commonly a shearing process. The highest level of torque sustained by the soil before shearing was noted and is recorded here in terms of soil shear strength.

4) The present topography of stable dunes and transgressive sand sheets was determined by precise levelling; and the stratigraphy and palaeotopography of the palaeosols identified, levelled, and mapped where exposures permitted in the vicinity of Coolatoo (fig. 7).

4. IMPACT OF CONTEMPORARY LAND USE

Compilation and examination of the available aerial photographs indicates that in the main study area south of Cantara (fig. 2) there has been extensive dune drift over the past 20 years.Inspection indicates substantial rates of advance are occuring opposite Dunn's Crossing and 32 mile Causeway. It is in these areas that the track network used by ORRV riders has increased most between 1954 and 1975; the sites also continuing to be major access points to the dunes.

These rates of dune drift cannot have been sustained for many decades, since extrapolation of this rate of drift back through time soon necessitates the landward edge of the mobile dunes being seaward of the present beach. There is no reason to believe such extensive deflation and/or landward migration has occurred. Early European descriptions from 1802 on suggest the presence of mobile sand on the seaward margins of the Peninsula—e.g. "All the coast along which we have run since yesterday is merely composed of sand dune inspiring melancholy and regret ..... the sea breaks all along the coast with extraordinary violence ...." Baudin's Diary 1902 (Cooper, 1952). Apart from the notoriously mobile Murray Mouth described by Thomson (1976), early maps and charts suggest the coastline was in basically the same position as today.

Consequently we conclude recent rates of dune advance are sufficient to cause genuine concern and there is good reason to suppose they are a consequence of recent unwise recreational land use.

5. DUNE DRIFT ASSOCIATED WITH EUROPEAN PASTORALISM

Whilst Cantara station continued as a pastoral homestead until recently, the Coolatoo homestead in the study area was abandoned at about the turn of the century. The area between the two was described by Todd (1856) as part of his survey for the construction of the Adelaide to Melbourne telegraph. The line of the route is illustrated in figure 3. Telegraph supports consisting of degraded wooden stumps or wooden poles with spindles or steel (Oppenheimer) poles used by the service have been relocated and mapped by Mr. J. A. Victory of Kingston, S. E. South Australia, and the author. Their precise location

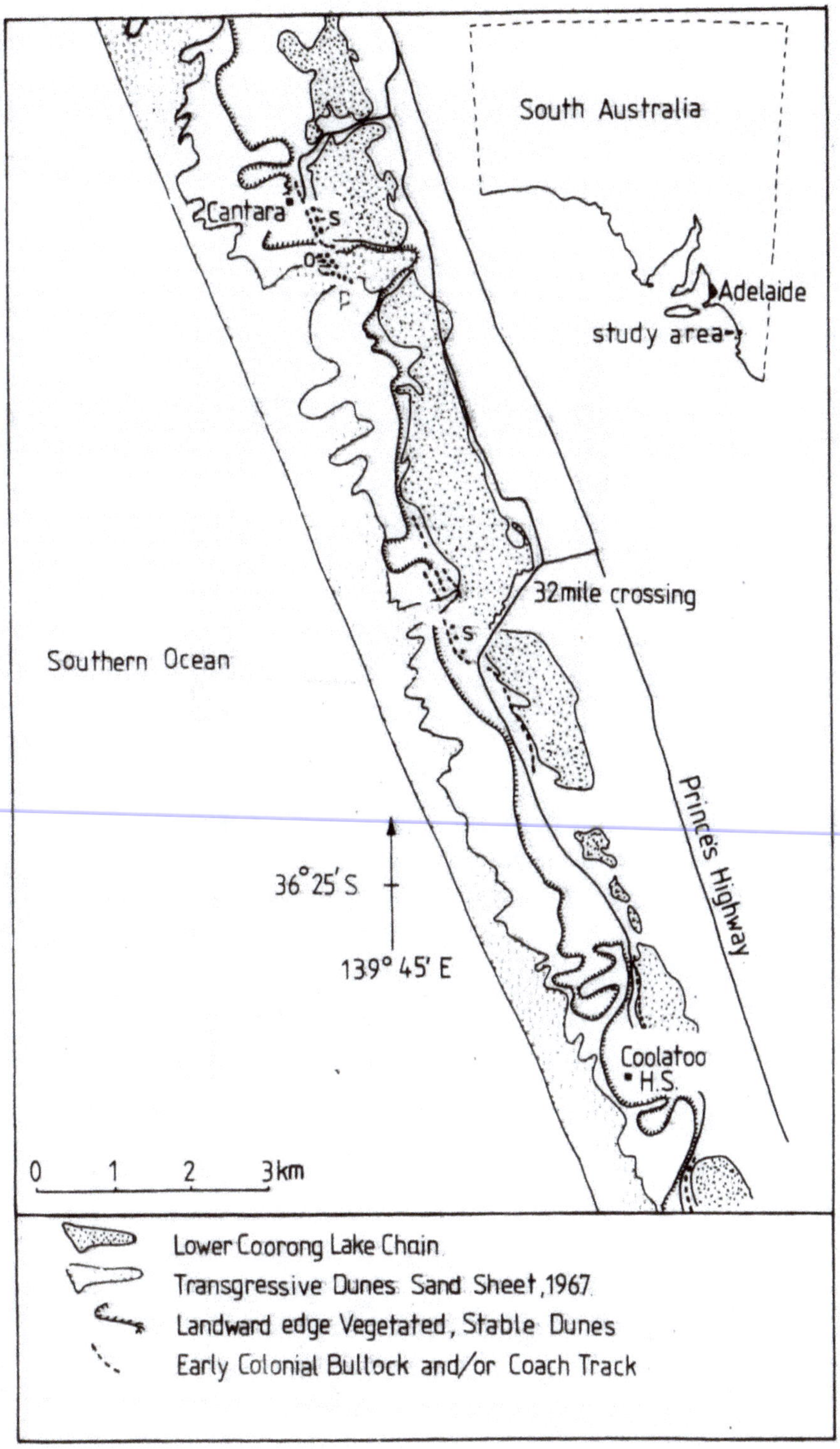

Figure 4. Location of *in situ* wooden stumps (. S), wooden poles with spindles (. p) or steel Oppenheimer poles (. O) used by the Adelaide-Melbourne telegraph line 1856-1929 south of Cantara.

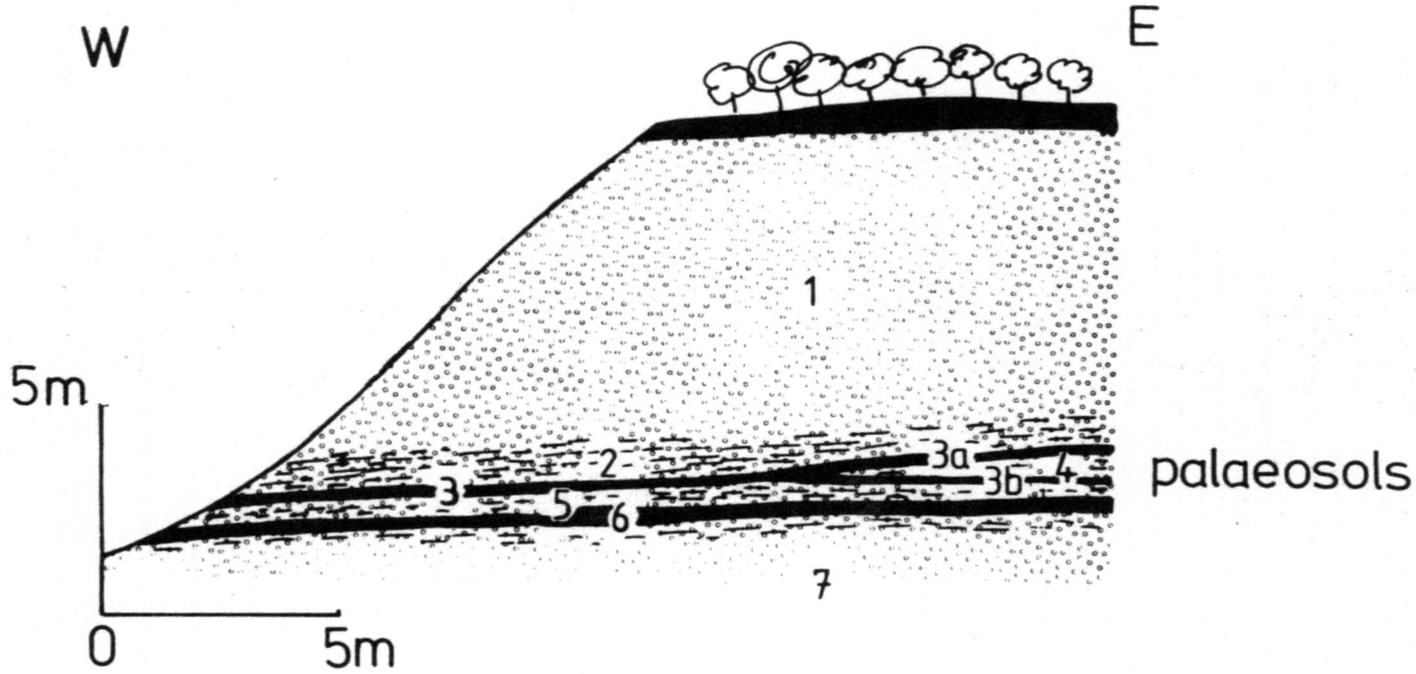

Fig. 5 Section through coastal dune sands near Coolatoo, showing dune sands and palaeosols with interbedded shell midden. For description see text.

in relation to current stable and mobile dunes in the study area is shown in figure 4. Todd's (1856) description of this area indicates active sand drift was only found immediately south of Cantara; no other active drifts were noted.

'From Tilley's crossing place keeping on the seaside of the Coorong, it is mostly a grassy plain, with a few rather high and steep sand hills running at right angles from the hummocks to the Coorong, evidently old drifts, but with one exception, which is still actively drifting, they are now well covered with grass."

A 15 m high drift can be traced up to, and probably buries the route at 32 mile causeway crossing. It is completely unmentioned and according to surviving members of the telegraph maintanance crew at Kingston, this drift could not be seen, if it existed at all, within 400 m of the route in 1929. However, this large active sand tongue was at approximately its present position in 1954, although then a narrow (100 m) arm of vegetated dune, 1 km seaward, separated it from the main sand sheet.

These data are inadequate to precisely specify the extent of actual dune advance associated with early European pastoralism. The 32 mile causeway crossing dune drift may be a legacy of this activity. Otherwise given the position of the dunes in 1954, and the records of extensive open drifts in the early and mid-nineteenth century, European pastoralism inevitably must be regarded as having induced much less dune advance than recent land usage.

The Aboriginal reference to "ants" could be taken literally. The area is ridden with large ants (see also Correll, 1963). Study of other landscape descriptions and interviews with the oldest residents of the area, sugeest that the rabbit infestation most notably brought about a change in the structure and composition of the vegetated dune vegetation from a savannah type with Casuarina tree spcies, to a continuous dense scrub dominated by the Coastal Wattle Acacia sophorae, rather than removal of all ground cover and exposure of open sand to deflation.

6. STABILITY AND INSTABILITY IN THE PRE-EUROPEAN DUNE SYSTEM

Precise accounts and radio-carbon dates of the pre-European aboriginal utilisation of the study area must await the publication of the Doctoral thesis of Luebbers (1978).

Dune Stratigraphy near Coolatoo

Within the dune sequence, pre-European aboriginal remains in this area are associated with a group of closely related palaeosols. The stratigraphic relationships are shown in figure 5, and given below.

Palaeosols and Midden near Coolatoo
(Grid ref. 3917, 59622, point A, Fig. 5)

Depth	Unit	Stratum
0-7 m	1	Residual mound of coastal, yellow carbonate dune sand, eroding on summit, N.S. & W. faces.
7-7.1 m	2	Organically stained, dune bedded carbonate sand - ? re-deposited soil from palaeosol below.

7.1-7.2	3	Shell midden in humus rich palaeosol, dip 50°W. Principally of Goolwa Cockle—Plebidonax deltoides, crab, some charcoal. This layer divides into two distinct organic A horizons of palaeosols 3a and 3b,
(7.05-7.10)	4	some 5 m S.E., separated by dune bedded yellow sand.
7.2-7.45	5	Humus enriched B horizon of palaeosol, becoming paler with depth.
7.45-7.8	6	Palaeosol, A horizon (0.05 m thick); rich in Plebidonx, gradual lightening in colour with depth.
7.8 on	7	Dune bedded yellow-brown carbonate sand.

Three periods of vegetation colonisation, dune stabilisation and soil development are noted, separated by brief periods of sand blow.

The palaeosols can be mapped, when exposed, as one unit throughout several tens of kilometres of the Peninsula. At many sites, sub-fossil scrub is revealed, still rooted in the soil. The midden material can be up to 0.35 m.thick and principally comprises flat bedded layers of thick, adult valves of the bivalve Plebidonax deltoides. This was used, with crab and fish as a food resource by the Aboriginal Tanganekald people, and obtained from the near-shore beach zone (Tindale, 1938). The palaeosol and midden surfaces are littered with hearths, hearthstones, scrapers, grind-stones, and ochre. The palaeosol is bound together by not only the midden layers and sub-fossil scrub roots, but also pipes of cemented calcium carbonate (calcrete), presumably originating as root exudates in this carbonate, over-rich environment. These pipes may be up to 0.7 m long and 0.04 m across.

Elsewhere in the Peninsula, broadly similar horizons in the dunes have been noted further north by Correll (1963) and Noye (1975), There is no information on their antiquity, or the time periods necessary for soil maturation and vegetation development to reach the level achieved. Radiocarbon dates in the range 6-8000 years B.P. from hearths and middens associated with paleosols in coastal dune sequences further south near Beachport reported by Tindale (1957) relate to older soils and sites which became buried by the equivalent of the yellow carbonate sands of Units 1-7 near Coolatoo; the Coolatoo palaeosols must be younger than these ages.

7. PALAEOSOLS AND DUNE STABILITY

The palaeosols and their contained cultural and biological remains shed light on the problem of stability in several ways.

Dune Stability on a Sub-Regional Scale.

The very existence of such well developed, widespread palaeosols, in the Peninsula, indicates that there have been significant periods of stability. The causes clearly operated at regional or sub-regional level. Either the scale and/or type of aboriginal land-firing changed in broadly synchronous fashion throughout the area; or alternatively natural oceanographic and/or climatic changes affected the dunes. The widespread character of the effect strongly suggests the latter. Although the effect obviously operated through greater vegetation cover and extensive soil development, there is insufficient data as yet to identify in detail the prime cause (s) of these changes.

Figure 6. Cumulative frequency curves showing variations in soil shear strength of soils, open sands, shell midden and palaeosols.

Sample Population	Material	Profile	Observations
1.1	Transgressive dune sheet adj. palaeosols	3	53
1.2	Main palaeosol with interbedded shell midden Unit 6.	3	72
1.3	"Recent" surface shell midden in dune swale	3	48
1.4	Open sand, transgressive dune sheet	3	52
1.5	Pre-European surface shell midden on palaeosols	3	47
1.6	Open sand in swale behind frontal dune ridge	3	67
1.7	Pre-European surface shell midden in thin palaeosol	2	47
1.8	Open sand-transgressive dune sheet	2	83
1.9	Modern soils on vegetated dune, open ground	1	23
2.0	Modern soil on vegetated dune beneath scrub	1	23

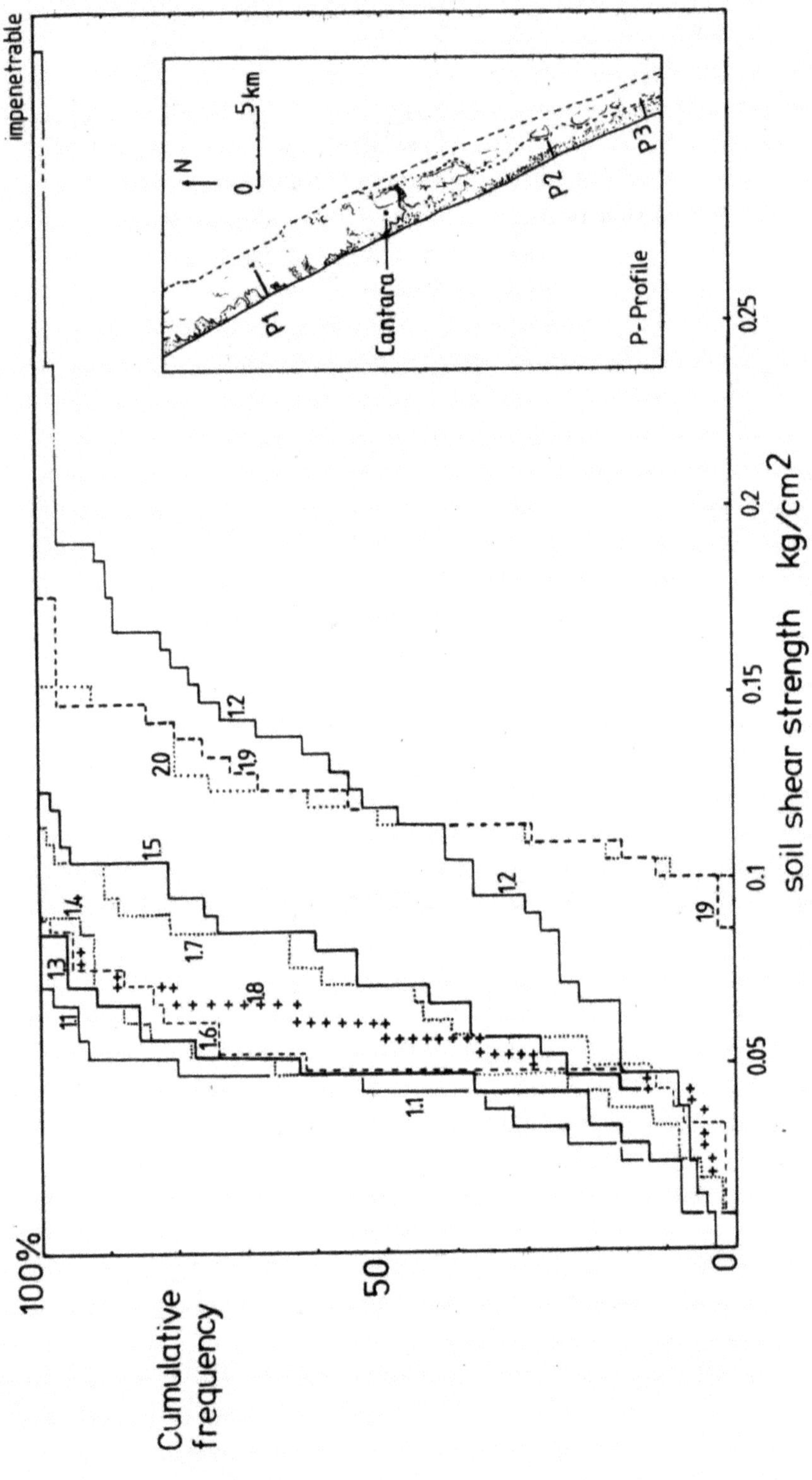

Cumulative frequency
100%
50
0
0.05
0.1
0.15
0.2
0.25
soil shear strength kg/cm2
impenetrable
1.1
1.2
1.3
1.4
1.5
1.6
1.7
1.8
1.9
2.0
N
0
5 km
P1
P2
P3
Cantara
P-Profile

Fig. 6

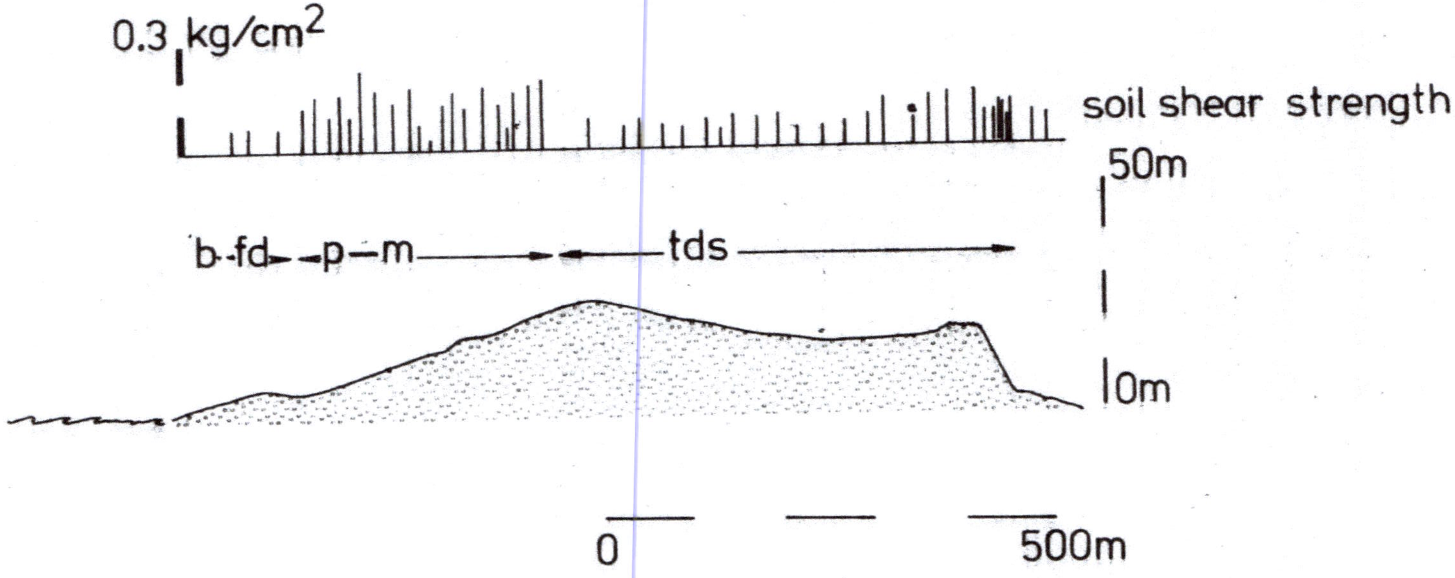

Figure 7. Outcrop of more resistant palaeosols identified by higher soil strength readings along profile 2 (figure 4): b= beach; fd = frontal dune ridge; p-m = palaeosol with interbedded shell midden; tds = transgressive dune sheet.

Possible Aboriginal Impact

The presence of interbedded midden and aboriginal remains in developing palaeosols (fig. 5) indicates aborigines could occupy the land without always inducing soil erosion and drift. Nevertheless, it is quite possible the small pulses of sand input represented by the sands and blown topsoil of Units 4 and 5 shown in figure 5, might well correspond to the effects of soil erosion consequent upon accidental or deliberate firing.

Palaeosols and Topographic Stability

Mapping and levelling indicated that deflation of dune sands tended to continue until the main palaeosol stratum was exposed. After this, sand mainly moved across the palaeotopography revealed.

The results of soil shear strength studies investigating the comparative resistance of contemporary soils, middens and palaeosols are shown in figure 6. The data were obtained by stratified sampling in the distinct units on 100 m.wide transects P1 - P2 - P3 located in figure 6. Each reading represents the mean of 3 readings taken in one square metre.

Open sands are seen to be relatively weak, with comparatively little spread in the strength range, samples 1.1 and 1.8 representing the minimum and maximum of the ranges. Recent midden (1.3) can be seen to be similarly unresistant. However, the pre-European middens in thin palaeosols (samples 1.5 and 1.7) can be seen to be notably stronger, but still weaker than modern dune soils. Here roots and other organic matter impart considerable strength. But it is in the thicker palaeosols that the notable strength of these features is clear.

The maximum strengths of some palaeosols could not be determined, and on 50% of occasions, these fossil soils were notably stronger than modern dune soils (samples 1.9 and 2.0).

The topographic consequences of this in higher resistance in the top 50% of measurements can be seen in figure 7. Deflation is occuring to the depth of the palaeosol throughout the seaward part of the profile, exposing the entire palaeotopography with its more resistant palaeosols. This topographic stability determined by prehistoric environmental change and human activity presumably represents the stability to which many of the users have been referring (see RVCC n.d.). It is also apparent in figure 6 that many exposures of the palaeosol are comparatively lacking in strength, and continual vehicle access can only disrupt them, and initiate the delfation of sub-palaeosol dune sands.

8. CONCLUSIONS

The data suggest that legacy effects from earlier European pastor lism and pre-European aboriginal activities are of secondary significance as causes of current sand drift problems compared to the impact of very recent land usage. In the prehistoric past, considerable changes in the stability of the coastal dunes have occurred at a scale which suggests naturally occuring environmental changes may have been the cause. The precise nature of these changes is as yet unspecified. Reports of apparent stability may relate to the exposure of comparatively stronger palaeosols and palaeotopography which are rich in archaeological remains.

9. ACKNOWLEDGEMENTS

The author acknowledges financial support from the Staff Research Funds of the Universities of Adelaide and Sheffield; to the South Australian Governments Department of Environment, National Parks and Wildlife, and Coastal Protection Board for considerable assistance; members of the Nature Conservation Society of South Australia Inc., for support in the field, assistance and information; to Mr. J. A. Victory of Kingston S.E., for his help in tracing the former Telegraph Route; Telecom Australia; Mr. Max Foale and his assistants for some of the Cartography and analyses; Dr. M. Arnold and the Department of Civil Engineering, University of Adelaide for the loan of equipment and discussion; and the following for their discussion of the sand drift problem and the area: Mr. K. Bardsley and Mr. D. Lock (Flinders University); Dr. J. R. Hails and Mr. D. Symon (University of Adelaide); Mr. R. P. B Bourman (Adelaide C.A.E.) and Dr. P. J. Lamb (Illinois Institute of Natural Resources).

BIBLIOGRAPHY

Bird, E. C. F., 1973. "Australian Caostal Barriers". Original Paper in Schwartz, M. L., 1973. Barrier Islands. Wiley.

Bourman, R. P., 1974. "Historic Geomorphic Change: Fleurieu Peninsula, South Australia". Proc. Int. Geogr. Union Reg. Conf. 8th New Zealand Geogr. Conf. Palmerston North. New Zealand Geog. Soc. (Inc.), pp. 289-297.

Cooper, H. M., 1952. French Exploration in South Australia, with special reference to Encounter Bay, Kangaroo Island, the Two Gulfs and Murat Bay, 1802 - 1803. Adelaide.

Correll, R. L., 1963. The application of Phytosociological Techniques to the Vegetation of the Younghusband Peninsula. Thesis submitted for the Honours Degree of Bachelor of Science, Department of Botany, University of Adelaide.

Easton, A. and Radok, R., 1970. Australia's Changing Mean Sea Level Surface. Horace Lamb Centre for Oceanographic Research Survey Paper no. 9. Flinders University of South Australia.

Gilbertson, D. D., 1977a. The Southern Coorong Lakes and the Lower Younghusband Peninsula - An Introduction, pp.1-18 in Gilbertson and Foale, 1977 op cit.

Gilbertson, D. D., 1977b. The Off-Road Use of Vehicles and Aspects of the Bio-Physical Systems of the Lower Coorong Region, pp. 71-91 in Gilbertson and Foale, 1977 op cit.

Gilbertson, D. D. and Foale, M. R., 1977. The Southern Coorong and Lower Younghusband Peninsula of South Australia. NCSSA. Adelaide. 121 pp.

Landsberg, H., 1956. The Orientation of Dunes in Britain and Denmark in Relation to Wind. Geogr. Jl. 122: 176-189.

Liddle, M. J. and Moore, K. G., 1974. The Microclimate of Sand Dune Tracks: the relative contributions of Vegetation Removal and Soil Compression. J. appl. Ecol. 11: 1057-1068.

Liddle, M. J. and Grieg-Smith, P., 1975a. A Survey of Tracks and Paths in a Sand Dune Ecosystem: I Soils. J. appl. Ecol. 12: 893-908.

Liddle, M. J. and Grieg-Smith, P., 1976. A Survey of Tracks and Paths in a Sand Dune System. II Vegetation. J. appl. Ecol. 12: 909-930.

Luebbers, R. A., 1978. Meals and Menus: A Study of Change in Prehistoric Coastal Settlement in South Australia. Doctoral Dissertation, Department of Prehistory, Research School of Pacific Studies, Australian National University, Canberra.

Noye, J., 1975. The Coorong (ed.) Department of Adult Education, University of Adelaide Publication no. 39.

Paton, J. B., 1977. Historical Notes on the Coorong. pp. 99-105 in Gilbertson and Foale op cit.

Ranwell, D. S., 1972. Ecology of Salt Marshes and Sand Dunes. Chapman and Hall.

RVCC (Recreational Vehicles Co-ordinating Council) n.d. Recommendations on Control and Management of Recreational Vehicles in South Australia. Report submitted to the South Australian Department of Environment.

Sprigg, R. C., 1952. The Geology of the South East Province of South Australia with special reference to Quaternary coastline migration and modern Beach Development. Geol. Surv. South Aust. Bull. no. 29.

Thom, B. G., 1974. Contemporary Coastal Erosion - Geologic or Historic? in The Impact of Human Activities on Coastal Zones. Australian Committee for Man and the Biosphere. Publ. no. 1, Aust. Gov. Publ. Service, Canberra, pp. 58-65.

Thomson, R., 1976. The Geomorphology of the Murray Valley in South Australia. Thesis for the Degree of Master of Arts, University of Adelaide, South Australia.

Tindale, N. B., 1938. "Prupe and Koromarange": A Legend of the Tanganekald, Coorong, South Australia. Trans. Roy. Soc. South Aust. 62(1): 18-24.

Tindale, N. B., 1957. A Dated Tartangan Site from Cape Martin, South East of South Australia. Trans. Roy. Soc. South Aust. 80: 109-123.

Todd, C., 1856. Proposed Connection with Melbourne by Electric Telegraph. Parliamentary Papers of South Australia, no. 11, Parliament House, Adelaide, Second Session, pp. 8-14.

Von der Borch, C. C., 1976. Stratigraphy and Formation of Holocene Dolomitic Carbonate Deposits of the Coorong Area, South Australia. J. Sedim. Petrol. 46(4): 952-966.

COASTAL RESOURCES AND THE CHANGING ENVIRONMENT

K. D. Thomas

Many of the marine animals which man exploits, and has exploited in the past, are ectothermic or cold-blooded forms whose body temperature is related to that of the sea-water. These animals include fish, crustaceans and molluscs. Even warm-blooded endotherms like seals, Cetacea and sea-birds are responsive to marked temperature change. In addition, many of the marine animals exploited by man have come from the littoral zone, where they are alternately exposed to the rigours of terrestrial life at low tide, and marine conditions at high tide, and are therefore susceptible to changes in both ecosystems. Of course, climatic change is only one aspect of environmental change which may influence marine life, especially in the littoral zone; local environmental changes may be of great importance as well as the effect or man. These are considered below.

The sources of Evidence

The nature and significance of various marine resources is considered in considerably greater detail elsewhere in this volume. Often, the most obvious evidence ocurs in the form of huge shell accumulations of various ages and these may be the only major sources of evidence about certain past cultures and their economy. The prevalence of low meat-yielding shellfish remains in such mounds may greatly exaggerate their importance in ancient diets (Meighan, 1969; Bailey, 1975). Such mounds usually contain the remains of other food resources—including crustacea, fish, birds, mammals and, sometimes, plants. Not all the material represents food; artefacts of stone may occur and often there is evidence that shells have been used as tools, ornaments etc.

With these shell mounds there is usually little doubt that the shells, and other materials, have been collected by man primarily for consumption. In other cases, such as where the remains are very scattered or occur in small quantities, the interpretation is less clear. Many sites of the Roman period in Britain include accumulations, of varying size, of oyster shells. From what we know of the Roman diet, we can be fairly sure that these usually represent food remains. Sometimes we may get evidence of utilisation; for example, a small collection of oyster (Ostrea edulis) shells near the 10th century A.D. Graveney boat was interpreted as food debris from the pattern of breakage on the valves, which was consistent with them being opened with a knife (Thomas, 1978). However, not all systematic damage on shells can be interpreted as being due to man; some species of crabs habitually prey upon gastropods by nipping off the top few whorls of the shell and pulling out the enclosed animal.

Shells with such patterns of damage may be common in beach deposits, and in deposits in caves near sea-level (which may also have been inhabited by man in the past). It is quite common for archaeologists, especially those working in the Mediterranean region, to ask if marine shells obtained from excavations were collected for food, or collected dead from a beach. It is often not easy to give an unequivocal answer unless large numbers of specimens are available. Beach-collected material is often very eroded ("beach-rolled") by wave action. However, it is possible to collect very fresh shells from beach deposits while living animals may have badly eroded shells. Some shells may be bored with small holes and the archaeologist may ask if this was done by man. There are a number of carnivorous gastropods which bore holes in the shells of their prey and then feed upon the shell contents through the hole. The hole is bored from the outside inwards with the aid of acidic secretions and the radular teeth and is a parallel-sided perforation. Man-made holes in shells, usually made for suspending them as ornaments, tend to be made from both inside and outside the shell with a pointed borer and are usually bi-conical in section. The occurrence of gastropod-bored shells on sites also indicates that they were not collected by man for food, since the occupants would have gone.

After death, bivalved molluscs leave behind the shell which usually separates into the individual valves when the proteinaceous ligament has decayed. When collected alive the valves of bivalves are, of course, always paired. Thus, a high frequency of matching valves (e.g. equal numbers of left and right hand valves with comparable size ranges) would indicate collection of the shells when their occupants were alive. Detailed analyses of valve-pairing have been made by Koike (1979).

Shells in the sea provide surfaces upon which various plants and animals settle from the plankton; some of these may form calcareous encrustrations on the surface of the shell. This is especially true of bryozoans, various marine worms (e.g. Spirorbis and Pomatoceros) and barnacles. When such epizoic forms enter a living shell they fail to form an encrustation and are often coated with a layer of calcium carbonate and thereby are incorporated into the shell of the mollusc. Sometimes they lie free inside the shell and become coated with many layers of calcium carbonate by the mollusc (this is how pearls arise in oysters). If we find shells with such epizoic forms encrusting the inside as well as the outside, we can be sure that the shell was empty for some time in the sea before it became incorporated into an archaeological context. It was not collected for food, unless in error.

Sometimes we can be fairly sure that the shells occur in an ancient beach deposit and the possibility that they were brought together by man does not arise. These ancient deposits are very useful in that they provide less biased evidence for climatic and other environmental factors. Material collected by man is often biased and may not include environmentally sensitive and diagnostic forms.

Aspects of Change

Patterns of change in past human economic activities are now well established in archaeology, at least in broad outline. Changes may be due to a complex of factors including cultural/technological change as a result of new innovations and perceptions. The nature of the change may be conditioned by the environment (i.e. not all things may be possible), but not necessarily determined by it. However, environmental change may necessitate some change in exploitation pattern.

Pleistocene climatic changes resulted in great changes in sea temperatures and sea levels. In the Holocene, maritime localities have undergone quite dramatic change as a result of glacial regression and eustatic rise of sea levels. In many parts of the world the early Holocene is marked by the presence of broadly-based hunter-gatherer economies (e.g. see Clarke, 1976) and in some cases the exploitation of marine resources appears to have been of some importance, at least on a seasonal basis. These patterns of exploitation are not limited in time to the early Holocene; some are much older, such as the South African shell accumulations in coastal caves (Voigt, 1975), while others are much younger and may even continue up to the present day (Bailey, 1975).

In this paper, I will be considering three main aspects of change:-

1. The effect of world-wide changes in climate upon marine ecosystems and human resources.
2. The effects of more localised and short-lived environmental changes.
3. Changing human subsistence patterns, including the effects of factors (1) and (2) above, and of other factors.

Obviously, it is not valid to treat these aspects as being wholly independent on the contrary they interact a great deal, as I hope to show below.

Major Climatic Shifts

The major climatic changes of the Pleistocene caused great changes in marine ecosystems—as are recorded in deep-sea sediments in various parts of the World's oceans. We are interested in the coastal, and particularly littoral, zones so the discussion here will focus upon these.

The climatic fluctuations in the Pleistocene were particularly associated with changes in temperature and the evidence from terrestrial deposits and deep-sea cores shows that both land and marine ecosystems were affected by them. Animals living in the littoral zone would be influenced by both marine and terrestrial conditions at various times during a single day. As a result, we have evidence of marked changes in the past in littoral faunas in relation to changes in sea temperature. For example Butzer (1975), in a discussion of sea-level changes and other phenomena on the Mediterranean island of Mallorca, points out that in some warmer phases of the Pleistocene molluscs now found in more southerly latitudes were present in the Mediterranean. These so-called "Senegalese faunas" include shallow-water and inter-tidal forms such as *Conus testudinarius*, *Strombus bubonius*, *Mytilus senegalensis* and *Arca plicata*. They are all absent from the Mediterranean at the present day. Voigt (1975) records certain littoral species from midden deposits in the Klasies River Mouth Caves in South Africa which show climatic change in the area. For example, the cold-water limpet *Patella granatina* was found only in the pre-Later Stone Age deposits.

The distribution and abundance of marine resources in relation to sea temperature is not reflected in simple latitudinal gradients of species diversity. Thorsen (1962), in a study of marine benthonic molluscs along the East Atlantic shelf from Gibraltar to Murmansk, showed that the highest numbers of species occurred in the Skagerrak—South-west Norway region, where Arctic cold-water forms occurred along with many warmer water Atlantic elements. This area represents an ecotone (Odum, 1971) between more Arctic and more temperate marine communities and, as Odum points out, the diversity and abundance of living forms may be greater in such transitional areas. They would certainly present good regions for human exploitation of resources.

Climatic change will influence the amount and distribution of sea-ice in northern latitudes at various times of the year. Simonsen (1972) has suggested that during the Late Stone Age in North Norway the human population practised nomadism, exploiting coastal sites in the winter (based on the occurrence of bones of Arctic coastal bird species on sites, as well as other evidence) and inland sites in the Spring, Summer and Autumn. In contrast, in areas where the sea freezes every winter (as on the northern Russian coast) coastal exploitation appears to have occurred in the summer only.

In the Holocene there have been marked climatic shifts at various times which, although of lesser intensity than the Pleistocene ones, have had an effect on faunas. Schledermann (1976) shows how changes in the amount of sea-ice over the last 1,000 years has had a great influence on the Thule Eskimo culture of Northern Canada. Warmer phases may have led to an increase in important food species, including bearded and harp seals and walrus, as well as white and bowhead whales and Narwhal. The ringed seal may have been favoured by colder conditions. In colder periods many of these species may have moved southwards and this could have influenced human settlement.

Climatic shifts also influence oceanic circulation and the position of the divergence and convergence zones of ocean currents. Ruddiman and Glover (1975) have demonstrated shifts in the North Atlantic circulation around 9,300 years BP on the basis of evidence from Foraminifera. They suggest that the oceanic and atmospheric circulations in the North Atlantic were influenced by the large Laurentide ice sheet still over eastern North America, and also by the cold Arctic Laurentide deglacial outflow. The effect of these changes, and of other comparable changes in ocean currents at other times, on both coastal faunas and man can only be surmised. It is clear that at the present day enormous biological productivity can occur in areas where ocean currents diverge or where they pull away from coasts (Tait, 1972; Boje and Tomczak, 1978). In such localities, upwelling of bottom sediments occurs and the seas become especially rich in nutrients. The resulting biomass of marine organisms can be enormous and even the adjacent terrestrial habitants can be greatly influenced. Important upwelling areas include the edge of the continental shelf off the coast of Peru where important fisheries occur and huge sea-bird populations exploit them; these latter have produced enormous deposits of guano on the adjacent coasts, which are important sources of phosphate fertilisers. The eastern Atlantic also has important upwelling areas associated with the Canary current. Here, again, we find very important fisheries (e.g. of sardines). In the past upwelling zones may have shifted as a result of changes in both ocean currents and winds and may have had an important impact on the ecology of the adjacent coasts.

Changes in the atmospheric circulation may have had other important effects but again these must remain a matter of speculation. First, wave action is very dependent upon winds and the intensity of wave action has a great influence on marine life—especially in the littoral and shallow-water zones. Exposure to wave action is a major ecological factor determining the distribution and abundance of certain inter-tidal species of animals and plants, especially on rocky shores (Lewis, 1964). Secondly, winds may influence man's capacity to exploit offshore resources. Leach (1979), in a study of the animal remains from the Washpool midden site in New Zealand, showed a change in resource exploitation from offshore fishing (probably requiring canoes) to inshore fishing around dissected rocky shorelines. This was probably not due to seasonal differences in exploitation but may be related to a climatic shift in the 14th Century A.D. Perhaps a change in wind conditions made offshore fishing from boats impossible; even today such fishing is only possible on a few days each summer.

Climatic changes associated with glaciation and deglaciation brought about other changes in coastal sites, particularly in the relative levels of land and sea. Regressions of the sea would lead to new territories for exploitation and new coastlines. Such events may occur during a glacial phase, although in Scandinavia there is evidence of marine transgressions during glaciation as a result of isostatic sinking of the land under the weight of the ice. Conversely, while in many areas the Holocene is a time of at first rapidly-rising and then relatively steady sea levels, in Scandinavia the isostatic rebound of the mantle has led to progressive uplift of the land and to a retreat of the coastline. This has led to a change in the location of settlements associated with maritime resource exploitation (e.g. Broadbent, 1978; Zvelebil, 1978). The present location of some archaeological sites can only be reasonably explained when past sea levels are taken into account. Broadbent (1978) gives a good example of this in the Lundfors stone age site in Västerbotten, Northern Sweden. This site is now some 78 metres above sea level and some 40 kilometres from the present coastline of the Gulf of Bothnia. At about 3,300 b.c. the Lundfors site was a coastal one, exploiting both maritime resources and animal resources in the coastal hinterland.

Rising Holocene sea-levels elsewhere in Europe have probably drowned and destroyed many coastal Mesolithic sites.

Local and Short Term Environmental Changes

Changes in sea-levels and isostatic uplift are aspects of local change which have global causes; these have already been discussed. In some areas changes in coastal aggradation may lead to a build-up of sands and silts which may alter the coastal morphology. When a shore becomes progressively more sandy, with rocks being buried, we may expect marked changes in the inter-tidal fauna and also the shallow-water sub-littoral fauna (for a discussion of the faunas of rocky and sandy shorelines, see Yonge, 1966). For the strandlooper sandy shores do not represent an ideal collecting ground because most of the inter-tidal animals burrow in the sand and are difficult to capture. Coastal aggradation may also lead to changes in settlement or in the function of settlements, as in the case of the Mediaeval harbours in southern England which became silted up. Changes in the pattern of coastal aggradation may be

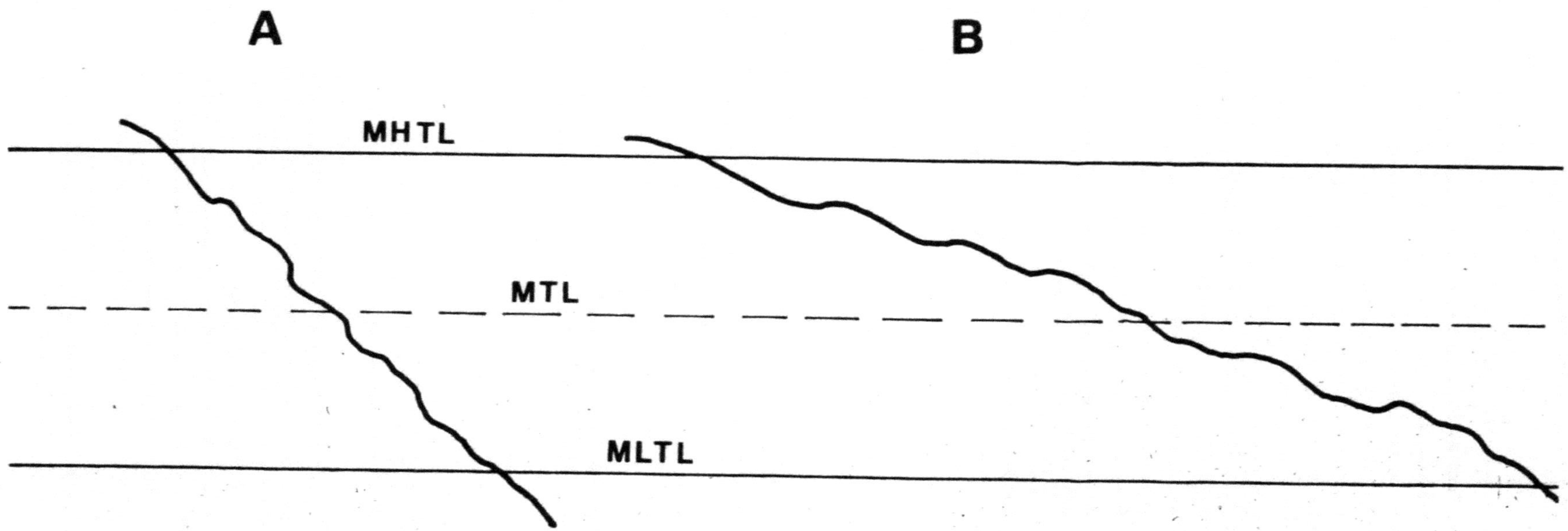

Fig. 1 Diagram showing the effect of coastal slope on the width of the inter-tidal zone for two rocky shores subject to the same vertical tidal range. Shore A is of steep morphology, with a narrow inter-tidal range, few rock pools and potentially subject to intensive wave battering. Shore B is of gently-sloping morphology, has a wide inter-tidal zone, many rock pools, and is washed by the tides.

Key: MHTL - mean high tide level
MTL - mean tide level
MLTL - mean low tide level

due to many complex factors including sea levels, currents, wave action and the sediment loads carried by run-off from the land.

Where many rivers run into a relatively land-locked sea we may get decreases in salinity. Deglaciation on the lands around the Baltic has led to an increased freshwater input into the Baltic Sea (Segerstråle, 1957). The subsequent decline in salinity has resulted in a fauna of low diversity which may not have been worth exploiting by human populations in the past.

On rocky shores local changes in coastal morphology in association with changes in the relative levels of land and sea, or by tectonic uplift or tilting, or by virtue of differential weathering of solid rocks, may lead to marked changes in the slopes of the shoreline. This has a great influence on the inter-tidal fauna and flora, and on the width of the inter-tidal zone (and hence the total biomass of inter-tidal life that can be sustained and exploited). This is shown in Fig. 1 for two hypothetical coastlines subject to the same levels of tidal rise and fall. Rocky shores with steep profiles are more exposed to the battering effects of waves and have a narrow inter-tidal zone. Their faunas are of low diversity, and the animals are very specialised and often small. Conversely, shores with gentle profiles are generally washed by the waves of the tidal flow, tend to have abundant pools remaining at low tide, and have a wide inter-tidal zone. They support a rich and diverse fauna and make good collecting grounds for strand-loopers.

Quite sudden changes in temperature due to local weather conditions may have a great impact on inter-tidal faunas but less so on sub-littoral faunas because sea-water temperatures are rather less susceptible to short-term change than air temperature. Animals which live in the inter-tidal zone tend to be sedentary (stenotopic) and usually have physiological adaptations which enable them to withstand the terrestrial conditions at low tide levels. Being ectotherms they are particularly sensitive to external temperatures. For example, Southward and Crisp (1954) record a great reduction in the abundance of the arctic-boreal barnacle Balanus balanoides in the Plymouth area over the period from 1934 (based on the data of Moore, 1936) to 1951. They attribute this to a slight rise in mean temperature over this period. Southward (1958) recorded lower lethal temperatures for a number of inter-tidal animals. He showed that sessile forms are tolerant of very low air temperatures (often to -10°C) while more active forms were less tolerant; these latter animals are probably capable of moving into warmer micro-habitats. Additionally, inter-tidal species living near the high-water level were more tolerant of low air temperatures than those species usually found near the mean tide or low-water levels. Obviously, the former species are exposed to air temperatures for much longer periods each day than those living further down the shore. In general inter-tidal stenotopes are more eurythermal and euryhaline than either their eurytopic or their sub-littoral counterparts.

Crisp (1965, and elsewhere) discusses the effects of the extremely cold winter of 1963 on the inter-tidal fauna of Britain. In general the southern elements were most affected (but not all of them) and the arctic-boreal species the least. Few species suffered marked changes in geographical distributions, but many of the non-sessile inter-tidal forms either retreated into sub-littoral habitats or were decimated. Slow-moving forms, especially many molluscs, were particularly affected by severe frosts. Some species of top shells

(Gibbula umbilicalis and Monodonta lineata) were reduced greatly in abundance on many rocky shores and became locally extinct on a few. It is clear that severe but short-term changes of weather can have a drastic impact on inter-tidal animals and may, in turn, influence the collecting activities of human populations exploiting these animals. This may affect both the amount of such animal food which can be collected as well as the range of species which may be available for exploitation. These affects on the fauna may persist for many years.

Changes in weather, or short-term climatic trends, may also influence sea-surface temperatures and, where there are systematic trends, even sub-surface temperatures. Beverton and Lee (1965) consider the possible hydrographic fluctuations which may have occurred in response to climatic variations in the North Atlantic Ocean over the last 100 years. There appears to have been a slight warming of sea temperatures and this is reflected by the increased abundance of cod (Gadus morhua) on the continental shelf of West Greenland since 1920. Fisheries statistics for the catches of cod by Greenlanders show a marked increase from some one thousand tons in 1920 to nearly 40,000 tons in 1960. Early cod fisheries in Greenland were dependent on migratory cod from Iceland but in the 1920s breeding populations of cod were established off Greenland and by the mid 1940s interchange with the Icelandic populations declined in parallel with reduced flow in ocean currents between these localities. In the North Sea, landings of sole (Solea solea) increased in the period 1946-1960 while those of haddock (Melanogrammus aeglefinus) declined. This again is consistent with warming; the sole is at its northern limit in the North Sea while haddock is at its southern limit. Changes of this sort in the past may have had an impact on populations dependent upon fishing, or upon certain methods of catching fish; in the above case there has been an increase in sole (a benthonic species) and a decrease in haddock (a pelagic form); these are species which require different fishing methods for their capture.

We have seen above that changes in the environment may have adverse or beneficial effects on some species, while others may be unaffected. In the littoral zone, however, the relative success of some species may have an impact on others as a result of competitive interactions. Two dominant ecological forces in the inter-tidal zone are the increased pressures of terrestrial life as one goes up the shore, and the increased pressures of competition (for food and, especially, space) and predation as one goes down the shore. Changes in these biological interactions, as a result of changes in the relative abundance of certain species, can have dramatic effects on the fortunes of other species. This is well illustrated by the work of Connell (1961) on the population biology of the barnacle Balanus balanoides in relation to competition with the barnacle Chthamalus stellatus and predation by Nucella lapillus.

Changing Human Subsistence Patterns

We have noted above that in some parts of the world there was a change in subsistence patterns in the Holocene towards a more broadly-based economy which sometimes involved the exploitation of marine resources (see, for example, Evans, 1969). In most of these areas this mode of life was later replaced by settled farming. During the phase of hunter-gathering subsistence

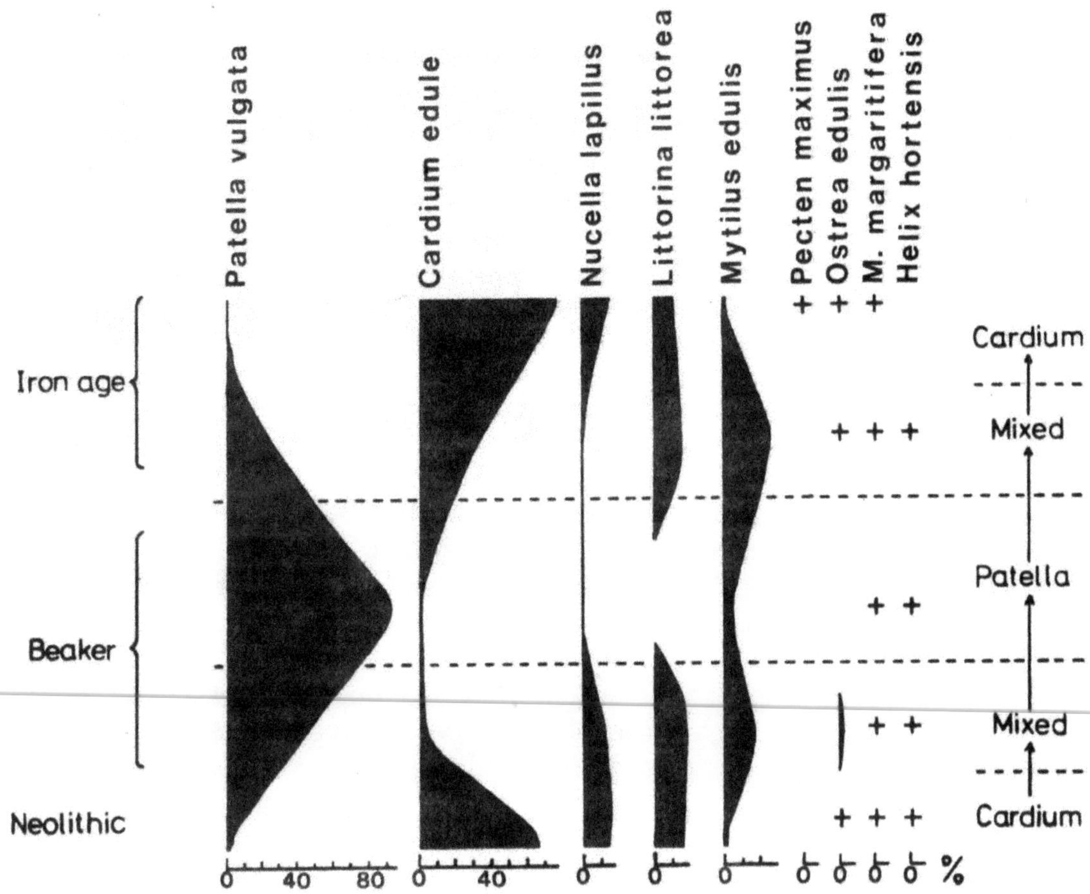

Fig. 2 Changes in the relative abundance of various food molluscs in a prehistoric shell midden at Northton, Isle of Harris, Scotland (modified from Evans 1969), with permission).

there must have been various factors causing change, both external ones and internal factors associated with human preference, technological advance, etc. Some of these will be considered below, insofar as they affect the occurrence of marine animal remains in archaeological contexts.

We have already seen how climate may influence geographic distributions of marine species and, hence, their availability for human exploitation. This is clearly shown in Voigt's study (1975) of the Klasies River Mouth Cave shell middens where Patella granatina occurs only in certain of the pre-Later Stone Age levels and was probably available only at those times because of lower temperatures. However, the results are not always so clear. Short-term changes of temperature may lead to a reduction in abundance of some marine forms (fish and inter-tidal animals) but we cannot be sure that a change in abundance of a species in a midden reflects this type of change, or some other change associated with patterns of human exploitation. Only where distinctive and diagnostic species occur in a deposit may we safely assume climatic change.

Changes in local marine and littoral environments may also lead to changes in the availability of species, which may be reflected in their abundance through a midden deposit. Evans (1969) has published the results of the analysis of a midden at Northton, Isle of Harris, in Scotland (Fig. 2). The dominant food species are limpets (Patella vulgata) and cockles (Cerastoderma edule); the former are most abundant in the Beaker period levels, the latter in the earlier Neolithic and later Iron Age levels. This may represent availability due to a change in environment, since Patella is common on rocky shores and Cerastoderma on sandy and muddy shores. This explanation, on its own, is not sufficient since the changes in the histogram for Cerastoderma are paralleled by changes in Nucella lapillus (dog whelk), Littorina littorea (winkle) and Mytilus edulis (mussel), all of which live on rocks on the shore. There may have been some environmental shift which was particularly detrimental to Patella; alternatively these changes may be due to factors associated with human exploitation (such as preferred areas of collecting, food preferences, or over-exploitation of particular species).

Voigt's study (1975) discussed above shows quite clearly how different zones on the shore may be exploited at different times; she found that in the earliest periods collecting was done over all zones but in the later period it was progressively restricted to the higher inter-tidal zones. The reasons for this are obscure—they may be due to external factors of the environment or to food collecting traditions or even the decline in importance of these inter-tidal food resources (at the present day in the Transkei shellfish are only eaten by females and very young boys).

This evidence for increasing collection in the highest zones of the shore is to be contrasted with the data collected by Fowler (quoted by Mellars, 1978) on the limpet shells from Mesolithic middens on Oronsay, Inner Hebrides. Here it was shown that the width/depth ratio of the collected shells was generally high (i.e. the shells were rather flat). In many modern inter-tidal environments limpet shells tend to become more conical as one progresses up the shore, so the data from the middens would seem to indicate collection from the lowest of the inter-tidal levels. This may have been due to dietary preferences for these and other animals in the lower zones or to earlier over-exploitation of limpets in the higher and more accessible zones of the shore.

Many middens include the remains of other animals as well as molluscs. Fish may be abundant and changes in the relative importance of these in comparison with the more sedentary inter-tidal resources, as well as changes in the representation of different fish species, may be related to environmental changes. Equally, however, they may be due to food preferences or to technological advances—such as the use of boats, fishing nets or traps, fish hooks and so on. Short-term changes may reflect seasonality of exploitation, or of availability, or both (see, for example, Leach, 1979). Sometimes only one major food resource will be represented; studies on the Oronsay middens (Mellars, 1978) have revealed that over 90% of the fish bones belong to the saithe or coalfish (Pollachius virens) which is very common in inshore waters.

We have already mentioned the possibility of the past over-exploitation of resources which may lead to a change in the representation of species in middens. Inter-tidal molluscs are rather stenotopic and therefore are easily collected (especially those living exposed on rocky shores). Over-exploitation by man is a distinct possibility and may be reflected in midden deposits by changes in species representation, changes in mean shell size or changes in the environments or littoral zones exploited. None of these changes can be easily interpreted as being the result of any single factor; however, some data may be very compelling. For example Evans (1978, 53-4) describes changes in a midden from Buckquoy, Orkney, in which a general decline in the frequency of limpets is matched by a fall in the mean size of the limpet shells. Unfortunately, data on the width/depth ratios are not presented so we do not have evidence for shifts in zones of collection. However, evidence for decline in abundance and size may together indicate over-exploitation. Okera (1976) has recorded that stocks of exploited (Senilia senilis (L), a shellfish common on muddy shores in Sierra Leone, tend to become dominated by the younger year-classes as a result of the collection of the larger shells.

Branch (1975) has recorded similar effects for recently exploited populations of the limpet Patella concolor in the Transkei. He showed that not only were the limpet shells in exploited populations much smaller than in other populations but the biomass (body, or flesh, weight) of the limpets was decreased by a proportionately even greater amount. A reduction in shell length of some 15 to 20 mm (as recorded for the size of the largest shells to be found in exploited populations over a period of five years) would result in a reduction of maximum body weight from 5.4 to 1.5 grammes. Thus, many more animals would have to be collected to obtain a comparable amount of meat. Branch further related the body weight of these limpets with the weights of the mature, and combined weights of the immature and spent gonads. The differences between these weights give an estimate of gametic output (i.e. roughly how many offspring could be produced per individual). A reduction of body weight from 5.4 to 1.5 grammes (following a reduction in shell size) would lead to a 90% reduction in gametic output. Clearly, selection of large-shelled limpets for human consumption has an enormous impact on the limpet populations not only in terms of immediate reduction in numbers but also in the biomass and future reproductive potential of the population.

Changes in the occurrence and the frequency of occurrence of species may be related to changes in dietary preferences. Voigt (1975) has shown for a modern midden in the Transkei that the composition of the midden closely resembles the composition of modern shellfish meals, but is quite different to the faunal composition of the local sea-shore. This may imply that the local shore has become depleted in the preferred species, or that other common species are ignored when the shellfish are collected. Preferences may change with time; a good example here is the exploitation of the oyster (*Ostrea edulis*) in Britain. This species was not particularly favoured in the prehistoric periods, judging by its low frequency in midden deposits, but was a great delicacy in the Roman period. Most Roman settlements in Britain yield large numbers of oyster shells. In the Middle Ages it fell from favour with the aristocracy and upper classes and was generally only consumed by the peasants. In recent years it has again become a delicacy and is, of course, very expensive.

Another aspect of changing exploitation patterns is the effect of seasonality. This may be due to seasonal occupation of sites, to seasonal exploitation of certain marine resources or to the seasonal availability of certain species. In a detailed and important study of a shell mound in Australia, Bailey (1975) showed that shellfish could not have been a major source of energy on an all-year-round basis. The data suggested either that shellfish formed a very minor part of the diet or that they were consumed in large numbers only at certain times of the years. Thus shell mounds may represent seasonally occupied sites or permanently occupied sites which have evidence for seasonal changes in resource exploitation. Studies on shell growth rings (e.g. Koike, 1975), on growth rings in fish otoliths and scales (e.g. Casteel, 1976) and oxygen isotope analyses of shells (Shackleton, 1969) may all cast light on seasonal exploitation. The analysis of relatively sedentary resources (such as shells) may yield evidence of seasonal occupation as may the occurrence of only certain types of seasonally available fish. On sites occupied permanently, changes in resources (e.g. changes in fish species available at different seasons) may indicate seasonal changes in exploitation (Leach, 1979).

Other facultative changes in exploitation may occur, but not necessarily on a seasonal basis. Branch (1975) describes how farmers in the Transkei have supplemented their diet with inter-tidal molluscs after bad crop failures in the early 1970s. In the later Holocene, exploitation of inter-tidal resources becomes of secondary importance after the arrival of agriculture but, in some cases, it persisted along with agriculture (e.g. the Ertebølle Culture, Troels-Smith, 1967). Perhaps inter-tidal animals were used as a supplement when crop yields or output by domesticated animals were low. Shellfish may also have been used for animal feed in some cases; for example, large numbers of limpet shells associated with a medieval farmstead near Beachy Head, Sussex, may represent feeding of limpets to pigs (Drewett, *et al.*, 1978).

In the later periods in many, but by no means all, regions some shellfish were viewed as luxury food items and were traded over wide areas (e.g. see Meighan, 1969); we have already seen this for the oyster in Roman Britain. Throughout prehistory and the historic periods marine shells have been prized as ornaments, currency, or for religious and ritual purposes as well as being used as tools and for other artefacts. Trade may be attested by the occurrence of marine shells on inland sites or in areas well outside their known geographic range (e.g. Jackson, 1917). Studies on the shell beads from the Neolithic site of Umm Dabaghiyah in north central Iraq (Thomas, unpublished data) showed evidence of trade with the Gulf because of the presence of various marine shells including those of *Paradentalium hexagonum* (Gould), *Cypraea moneta* Linn. and *C. annulus* Linn.

Finally, man may have had an effect on marine resources by the accidental introduction of new species, or by the purposeful culturing of others. Oysters are, yet again, a good example of both these aspects. Demands for oysters in Britain at the present day outstrip the rates at which beds of the native osyter (*Ostrea edulis*) can meet the supply; most oysters consumed in restaurants are carefully cultured in oyster farms and are the Portuguese oyster (*Crassostrea angulata*). Natural oyster beds have long been commercially exhausted by over-exploitation, and were further decimated by introduced competitors and predators (which can also be a menace in man-made oyster beds). Attempts in the latter part of the 19th Century to introduce the American oyster (*Crassostrea virginica*) into Britain led to the accidental introduction of two serious pests of oyster beds; the slipper limpets (*Crepidula fornicata*), which competes successfully with oysters for food and space, and the oyster-drill (*Urosalpinx cinerea*), which is a gastropod which preys upon limpets (Yonge, 1966).

Summary

This essay has covered some wide-ranging aspects of environmental change in relation to the availability of coastal resources, especially of those in the littoral zone. The discussion has mostly been concerned with the Mollusca as they are among the most commonly preserved and most easily identified of the marine resources exploited by man. Their numerical dominance vastly over-estimated their dietary significance as most species have a low meat: shell weight ratio. I have considered in some detail the evidence which may allow us to conclude that a shell was collected when its occupant was alive and therefore probably represents food debris. In many cases, such as where we have huge man-made shell mounds, the evidence of exploitation for food is unequivocal; in other cases we may not be so sure.

Aspects of environmental change have been considered in terms of world-wide climatic change, of localised environmental and short-term changes, and of changing human subsistence patterns and their possible impact upon maritime resources. The emphasis throughout has been on the effect of these various factors upon the representation of marine animals in shell accumulations and in other archaeological contexts. World-wide climatic changes were shown to have an important impact on the geographical distribution of marine species, not only in relation to air and sea temperatures but also through other factors such as the amount of sea-ice, the circulation of ocean currents, and the prevailing wind directions in coastal areas. Glacial advances and retreats have led to changes in sea-levels and to isostatic changes in the land; these effects have probably had an impact of human settlement patterns. Additionally, Holocene rises in sea level have obliterated many coastal archaeological sites.

Local environmental changes are often complex and correlated with global climatic changes. Changes in coastal aggradation, the slopes of shorelines, exposure, salinity, and local weather conditions may have profound effects upon the inter-tidal biota and will influence the availability of various resources for human exploitation. The interpretation of the significance of changes in the abundance, and even occurrence, of species through a midden deposit is made even more difficult by the effect of man's activities. Selective collecting in different habitats or shore zones, over-exploitation of resources, changes in dietary preferences, and seasonal exploitation are all factors which can influence the composition of midden deposits.

BIBLIOGRAPHY

Bailey, G. N., 1975. The role of molluscs in coastal economies: The results of midden analysis in Australia. J. Arch. Sci., 2: 45-62.

Beverton, R. J. H. and Lee, A. J., 1965. Hydrographic fluctuations in the North Atlantic Ocean and some biological consequences. In: C. G. Johnson and L. P. Smith (eds.), The Biological Significance of Climatic Change in Britain, pp. 79-107. Symp. Inst. Biol. No. 14, Academic Press, London.

Boje, R. and Tomczak, M. (eds.), 1978. Upwelling Ecosystems. Springer-Verlag, Berlin.

Branch, G. M., 1975. Notes on the ecology of Patella concolor and Cellana capensis, and the effects of human consumption on limpet populations. Zoologica Africana, 10: 75-85.

Broadbent, N. D., 1978. Prehistoric settlement in northern Sweden: A brief survey and a case study. In: P. Mellars (ed.), The Early Post-glacial Settlement of Northern Europe, pp. 177-204. Duckworth, London.

Butzer, K. W., 1975. Pleistocene littoral-sedimentary cycles of the Mediterranean basin: A Mallorquin view. In: K. W. Butzer and G. Isaac (eds.), After the Australopithecines, pp. 25-71, Mouton, The Hague.

Casteel, R. W., 1976. Fish Remains in Archaeology. Academic Press, London.

Clarke, D. L., 1976. Mesolithic Europe: The economic basis. In: G. de G. Sieveking, I. H. Longworth and K. E. Wilson (eds.), Problems in Economic and Social Archaeology, pp. 449-481. Duckworth, London.

Connell, J. H., 1961. Effects of competition, predation by Thais lapillus, and other factors on natural populations of the barnacle Balanus balanoides. Ecol. Monogr. 31: 61-104.

Crisp, D. J., 1965. Observations on the effects of climate and weather on marine communities. In: C. G. Johnson and L.P. Smith, (eds.), The Biological Significance of Climatic Changes in Britain, pp. 63-77. Symp. Inst. Biol. No. 14, Academic Press, London.

Drewett, P., Bedwin, O., Freke, D. and Rudling, D., 1978. Rescue Archaeology in Sussex, 1977. A fourth progress report on the Sussex Archaeological Field Unit. Bull. Inst. Archaeol. Univ. London, No. 15: 49-72.

Evans, J. G., 1969. The exploitation of molluscs. In: P. J. Ucko and G. W. Dimbleby (eds.), The Domestication and Exploitation of Plants and Animals, pp. 479-484. Duckworth, London.

Evans, J. G., 1978. Introduction to Environmental Archaeology. Elek, London.

Jackson, J. W., 1917. Shells as Evidence of the Migration of Early Culture. Manchester University Press, Manchester.

Koike, H., 1975. The use of daily and annual growth lines of the clam Meretrix lusoria in estimating seasons of Jomon Period shell gathering. In: R. P. Suggate, and M. M. Cresswell (eds.), Quaternary Studies, pp. 189-193. Royal Society of New Zealand, Wellington.

Koike, H., 1979. Seasonal dating and the valve-pairing technique in shell-midden analysis. J. Arch. Sci., 6: 63-74.

Leach, B. F., 1979. Fish and crayfish from the Washpool midden site, New Zealand: Their use in determining season of occupation and pre-historic fishing methods. J. Arch. Sci., 6: 109-126.

Lewis, J. R., 1964. The Ecology of Rocky Shores. English Universities Press, London.

Meighan, C. W., 1969. Molluscs as food remains in archaeological sites. In: D. R. Brothwell and E. S. Higgs (eds.), Science in Archaeology, pp. 415-422. Thames and Hudson, London.

Mellars, P., 1978. Excavation and economic analysis of Mesolithic shell middens on the Island of Oronsay (Inner Hebrides). In: P. Mellars (ed.), The Early Postglacial Settlement of Northern Europe, pp. 371-396. Duckworth, London.

Moore, H. B., 1936. The biology of Balanus balanoides. V. Distribution in the Plymouth area. J. Mar. Biol. Ass. U.K., 20: 701-706.

Odum, E. P., 1971. Fundamentals of Ecology, Ed. 3. Saunders, London.

Okera, W., 1976. Observations on some population parameters of exploited stocks of Senilia senilis (= Arca senilis) in Sierra Leone. Marine Biology, 38: 217-229.

Ruddiman, W. F. and Glover, L. K., 1975. Subpolar North Atlantic circulation at 9,300 years BP: Faunal evidence. Quaternary Research, 5: 361-389.

Schledermann, P., 1976. The effect of climatic/ecological changes on the style of Thule Culture winter dwellings. Arctic and Alpine Res., 8: 37-47.

Segerstråle, S. A., 1957. Treatise on Marine Ecology. Geol. Soc. Am. Mem., 67: Vol. 1.

Shackleton, N., 1969. Marine Mollusca in Archaeology. In: D. R. Brothwell and E. S. Higgs (eds.), Science in Archaeology, pp. 407-414. Thames and Hudson, London.

Simonsen, P., 1972. The transition from food-gathering to pastoralism in northern Scandinavia and its impact on settlement patterns. In: P. J. Ucko, R. Tringham and G. W. Dimbleby (eds.), Man, Settlement and Urbanism, pp. 187-192. Duckworth, London.

Southward, A. J., 1958. Note on the temperature tolerances of some inter-tidal animals in relation to environmental temperatures and geographical distribution. J. Mar. Biol. Ass. U.K., 37: 49-66.

Southward, A. J., and Crisp, D. J., 1954. Recent changes in the distribution of the inter-tidal barnacles Chthamalus stellatus (Poli) and Balanus balanoides (L.) in the British Isles. J. Anim. Ecol., 3: 163-177.

Tait, R. V., 1972. Elements of Marine Biology. 2nd ed. Butterworths, London.

Thomas, K. D., 1978. The Molluscs. In: V. Fenwick (ed.) The Graveney Boat: A Tenth Century find from Kent, pp. 155-160. B.A.R. British Series, No. 53.

Thomas, K. D., n.d. The shell and stone beads from Umm Dabaghiyah, Iraq. Unpublished m.s.

Thorson, G., 1962. The distribution of benthic marine Mollusca along the North-east Atlantic shelf from Gibraltar to Murmansk. Proc. First European Malacological Congress, 5-23.

Troels-Smith, J., 1967. The Ertebølle Culture and its background. Palaeohistoria, 12: 505-528.

Voigt, E. A., 1975. Studies of marine Mollusca from archaeological sites: Dietary preferences, environmental reconstructions and ethnological parallels. In: A. T. Clason (ed.), Archaeozoological Studies, pp. 87-98. North Holland, Amsterdam.

Yonge, C. M., 1966. The Sea Shore. Collins, London.

Zvelebil, M., 1978. Subsistence and settlement in the north-eastern Baltic. In: P. Mellars (ed.), The Early Postglacial Settlement of Northern Europe, pp. 205-241. Duckworth, London.

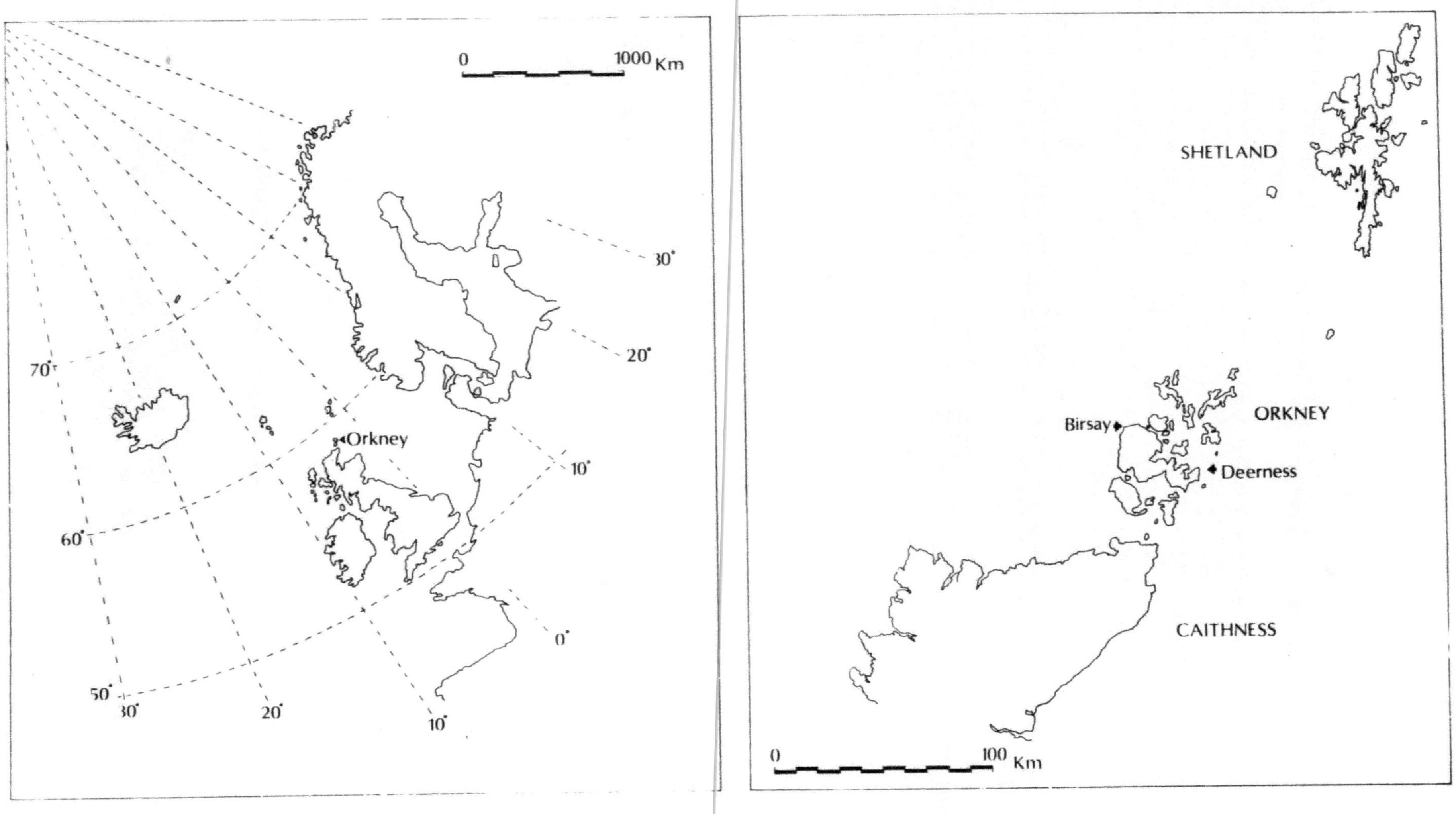

Figure 1. Location maps for Birsay, Orkney.

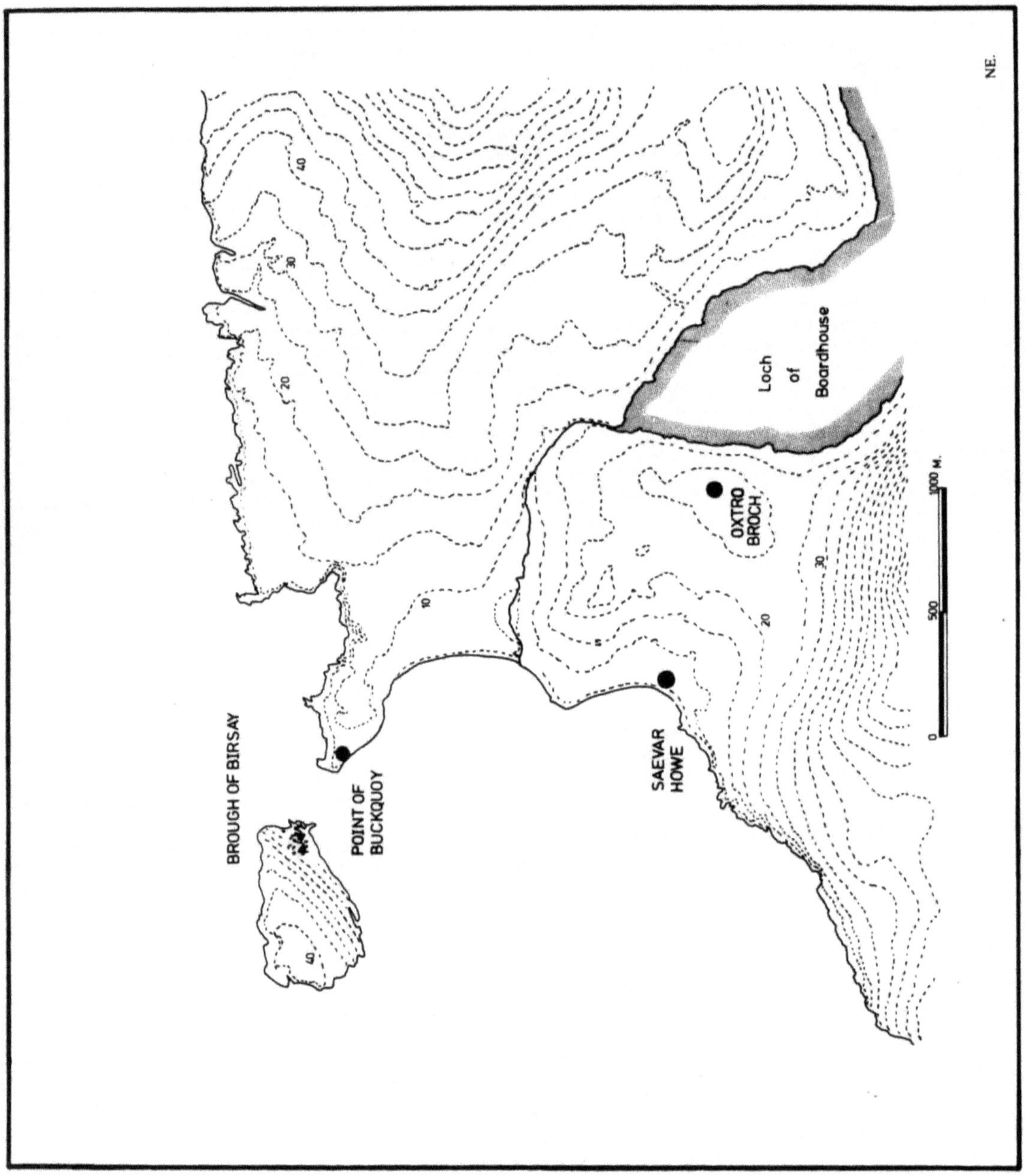

<u>Figure 2.</u> Contour map of Birsay region with location of early excavations.

THE BIRSAY BAY PROJECT

Preliminary investigations into the past exploitation of the coastal environment at Birsay, Mainland, Orkney

A. M. Donaldson, C. D. Morris* and D. J. Rackham

The Bay of Birsay is situated in the extreme north-west of Mainland, the largest island of the Orkney group, and is exposed to the North Atlantic except for the protection afforded by the Brough of Birsay, a tidal island forming, with the Point of Buckquoy, the northern edge of the bay (Figs. 1 & 2). The coast to the north-east and south has high cliffs and the bay forms the most northerly of the only three sheltered landfalls on the west coast of the island. The landward area of the bay is relatively flat with high ground rising up to 100 m to the south and west. Boardhouse Burn flows into the bay two hundred metres north of the Point of Snushan and is fed by the Loch of Boardhouse one kilometre to the south-east (Fig. 2).

The coastal fringe of the mainland area of the bay is characterised by deposits of wind-blown shell-sand capped by a grass sward. These deposits vary in thickness with the fall and rise of the underlying drfit deposits, sometimes reaching depths of three metres or more. The wind-blown sand deposits are underlain by boulder clay which lies on the Stromness Flags of the Middle Old Red Sandstone (Mykura, 1975). These gently inclined sandstones are visible around virtually the whole of the bay at low water and form a considerable hazard for any but small boats trying to beach. The sandstone, below the recent drift deposits, is present in low cliffs for parts of the bay and rises sharply to the south and north, forming cliffs over seventy metres high which act as a haven for thousands of sea birds.

The earliest archaeological horizon on the bay is represented by a buried soil profile forming a ground surface, possibly truncated, at the junction between the boulder clay and the shell sand around much of the bay. A monolith of this profile, collected from the west side of the Point of Buckquoy, suggested upon analysis a buried humic horizon within the profile, 0.1 m below the base of blown-sand incorporation in the soil. Superficial disturbance, possibly cultivation, could account for the soil sequence (Limbrey, pers. comm.). The lower part of the blown sand has become mixed with the underlying soil whether by worm action or 'continued cultivation' during the early part of the sand blow is not known although Spencer (Evans and Spencer, 1979) has suggested the former. This section was sampled for terrestrial molluscs in 1971 (Evans and Spencer, 1979) and the fauna from the buried soil indicates a woodland environment. Evans (1979) comments on the absence of open country species and has presented the view that the fauna is "indicative of the presence of true woodland prior to sand deposition".

* C. D. Morris is preparing an archaeological review of this Birsay work.

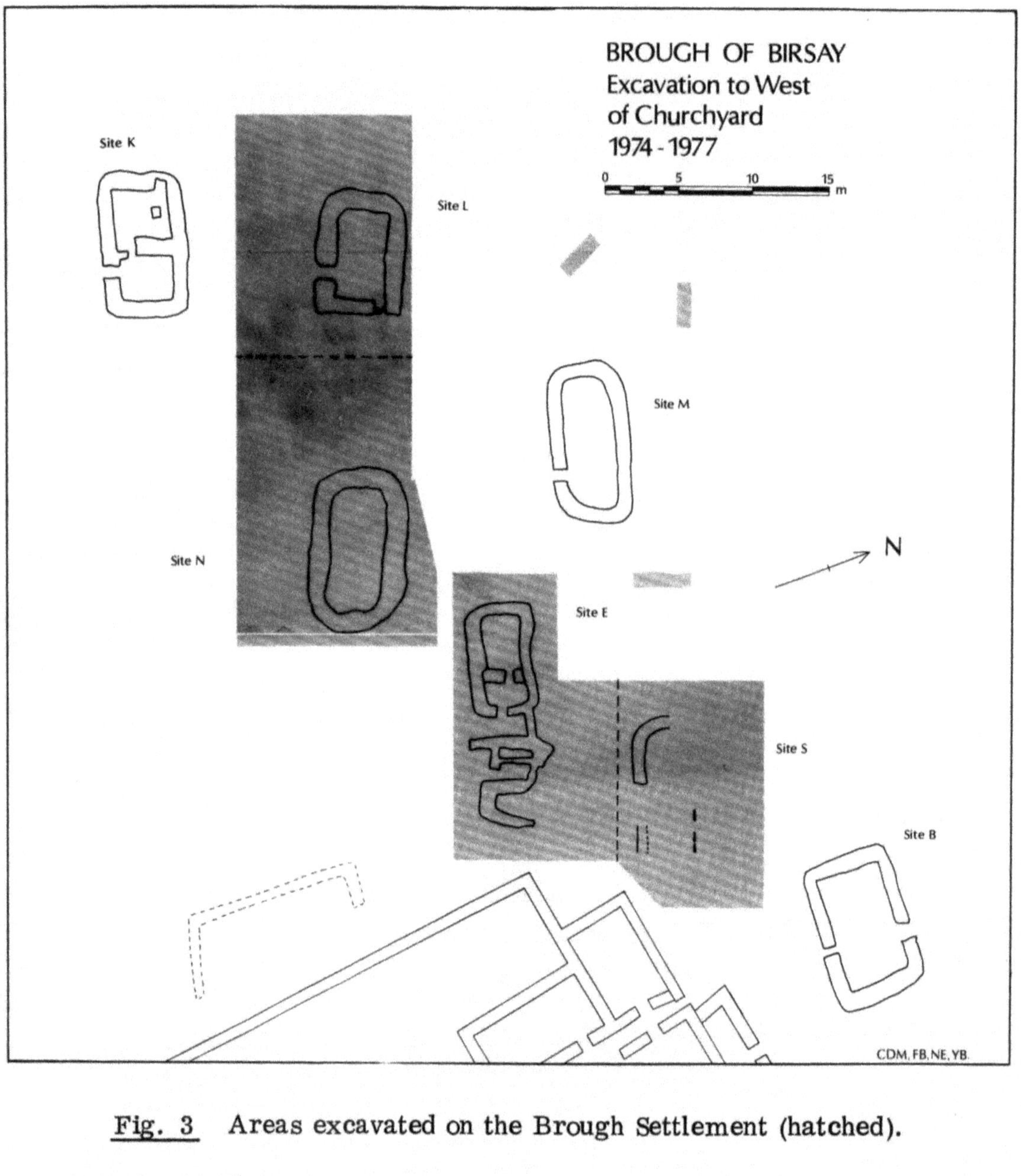

Fig. 3 Areas excavated on the Brough Settlement (hatched).

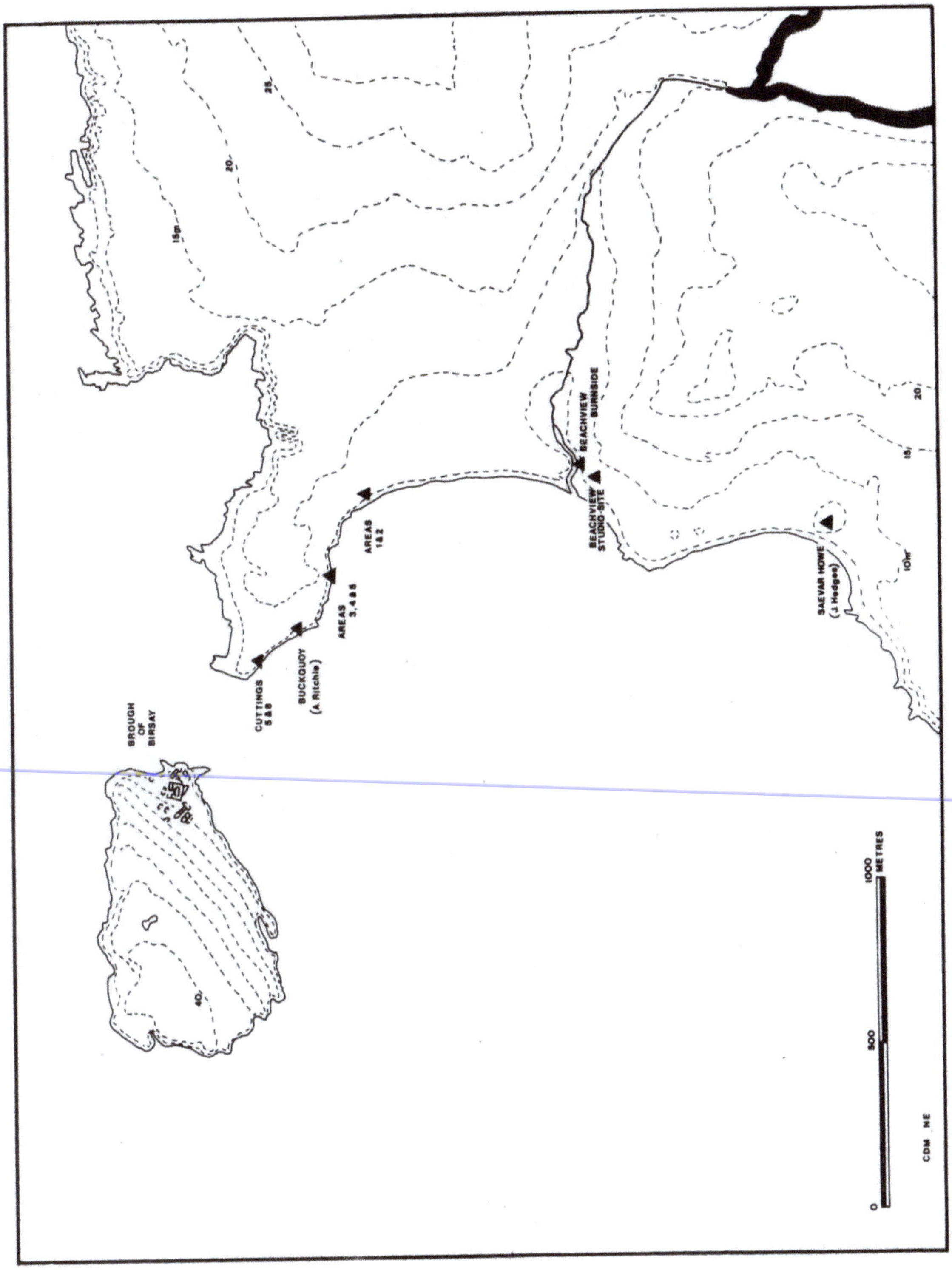

Figure 4. Birsay Bay. Recent excavations (since 1970).

Several pollen diagrams from mainland Orkney (e.g. Keatinge and Dickson, 1979; Davidson et al. 1976; Moar, 1969) show very low percentages of tree and shrub pollen throughout the prehistoric and historic periods, and most authors suggest that Orkney was never forested or wooded in the conventional sense. These diagrams however tend to be from rather exposed areas with large expanses of peat and the possibility of local shrub woodland in the more sheltered areas has often been overlooked. A short diagram from a small peat deposit in Deerness (Donaldson, unpublished) shows a period with consistently 28-50% tree pollen and 14-22% shrub pollen. Birch, willow, hazel and pine form a considerable proportion of the pollen sum and must surely indicate local native woodland or scrub during some part of the Post-glacial. Relict birch woodland occurs today in parts of Hoy (Prentice and Prentice, 1975) and there is no doubt that even small pockets of woodland would have affected and influenced local settlement and economy

Archaeological Investigations

Archaeological investigation in the area of Birsay began in the nineteenth century, with work carried out at Saverough (Saevar Howe) a large sand mound on the coast, Oxtro Broch near the western edge of the Loch of Boardhouse and the Brough of Birsay (see Fig. 2). When this latter site came into Guardianship in the 1930s, a renewed excavation programme began by the then Office of Works, and, following interruption due to the Second World War, was resumed from 1956 to 1964. In the meantime investigations had also taken place at the Knowe of Buckquoy and on the tip of the Point of Buckquoy (see Fig. 2). In 1970-71 Dr Anna Ritchie undertook the excavation of a site threatened by coastal erosion in the northern part of the bay, and this has recently been published (Ritchie, 1979), and in 1973 a small excavation was undertaken on the Brough of Birsay by Mrs C. L. Curle in association with Durham University. This work was completed in 1974 and a large scale excavation programme initiated on the Brough.[1] In 1976 monitoring began of threats to archaeological sites around the bay from coastal erosion and other agencies, with some salvage excavation. In 1978 excavations began on a range of sites on the mainland side of the bay,[2] and in 1977 J. W. Hedges undertook a further excavation at Saevar Howe (see Fig. 4). Work on both the Brough and the mainland sites is continuing.

In the current excavations on the Bay of Birsay, the archaeological research has been concentrated on the pre-Norse and Norse periods, but the known sites span the period of time from the Bronze Age onwards. In 1978, trial work at Cuttings 5 and 6 (Fig. 4) on putative boat-nousts, revealed a build-up below. Indeed this site, on the west side of the Point of Buckquoy, overlying the soil profile described above, contains a stone feature visible in the cliff section with which a midden is associated. This is the earliest archaeological feature known in the bay, and appears to be contemporary with the onset of sand-blow in the area. Although excavation was not possible a sample of 80 kg of the midden was taken. This produced an abundance of biological remains including carbonised naked barley that has since been radiocarbon dated to 3260 ± 180 bp (GU-1222). The inhabitants of this period exploited a wide range of animals typical of the coastal situation, and in particular the fish assemblage suggestes that the major fishing effort was on the shore oı close inshore. Several species could have been caught at low tide

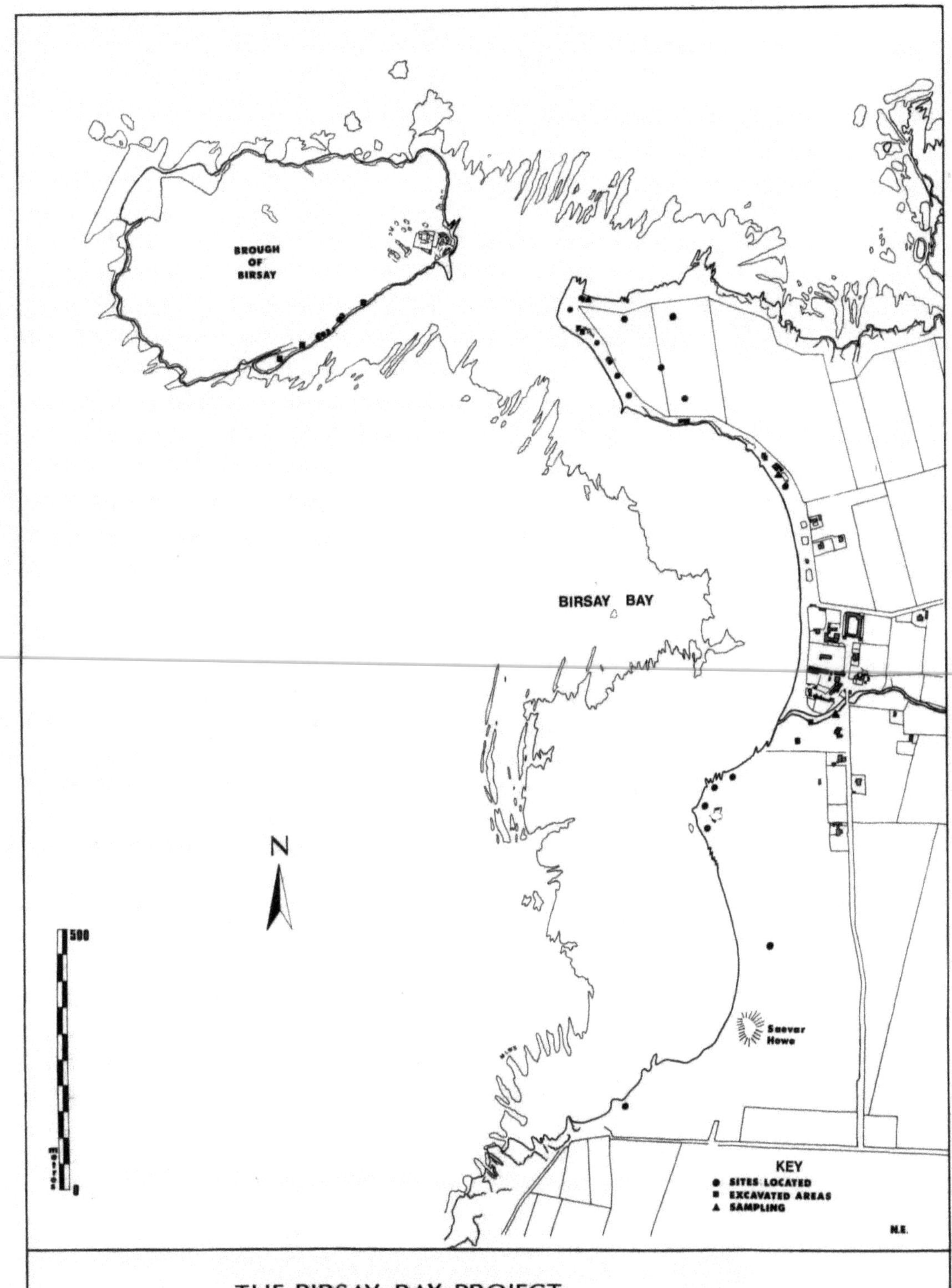

Fig. 5 Known archaeological sites on the Bay of Birsay.

in rock pools (Alvey et al., unpubl. report). Falling within the chronology of the Bronze Age, for which few environmental data are available for Orkney, this small sample interestingly suggests a major dependence upon the local environment for food. The mammalian fauna is dominated by red deer, of which at least four individuals are present, and the wild birds, fish and marine molluscs form a major part of the fauna, which originally led to the hypothesis that the site was of Neolithic date. Only two ox bones were identified, but the presence of over two hundred carbonised grains of naked barley, and the possible cultivation of the soil noted above, do indicate agriculture as a significant part of the economy of the site.

The land snail fauna associated with the shell-sand covering these deposits (Evans and Spencer, 1979) contains the grassland species Vallonia excentrica and may indicate the clearance of the woodland or scrub; whether due to the action of sand-blow or man is undetermined. The accumulation of sand dunes over this area must represent a major change in the coastal environment at this time, probably reducing the coastal agricultural zone.

Neither of the excavations at Saevar Howe and Oxtro have left any details of the environmental remains, but the identification of Oxtro as a broch with later phases and the finding of probable early Christian (i.e. pre-Norse) graves in the sand covering a major structure at Saevar Howe indicate habitation in the area during the first millennium B.C. and the early part of the first millennium A.D. These are clearly major multi-period sites, and a third has been identified in the current work at Beachview (Burnside site 2) where a sequence of major structures underlies deposits broadly dated to the Norse period. No work has yet been under taken on these structures, and therefore no environmental data are yet available, but in time it may be hoped that material deriving from contexts of the Iron Age extending through to the immediately pre-Norse period may be analysed.

Investigations this century have been largely concerned with the last two millennia, particularly the pre-Norse (also termed Early Christian and Pictish, see Ritchie, 1975) and Norse remains at a number of sites around the bay. The bay and adjacent land has a wealth of archaeological remains, many of which date from the Viking and later Norse /Medieval periods. Indeed one is continually reminded of the Norse influence in the many place names of Old Norse origin (Marwick, 1970). The Orkneyinga Saga indicates that in the eleventh and twelfth centuries the bay was the home of the Earls of Orkney: for instance Earl Thorfinn "had his permanent residence at Birsay, where he built and dedicated to Christ a fine minster, the seat of the first bishop of Orkney" (Palsson and Edwards, 1978, 71). Parts of the walls of the later Stewart Earls' Palace still stand to over twenty feet in height in the village, and this area, with the parish church, has been claimed by some as the focus of settlement in these periods (Lamb, 1975).

At this stage of archaeological analysis it is not possible in all cases to clearly distinguish the pre-Norse and Norse material: some excavations are on-going, others await radiocarbon determinations and other work. Equally, the analysis of the environmental data from recent excavations is incomplete and on-going. Environmental material from excavations conducted before the present decade are either lost or unreported, and apart from information relating to the structures, age and the small finds themselves, few data are left to us. Of material from excavations by other archaeologists, that from Buckquoy is now published (Ritchie, 1979), but although recent excavations by Hedges (Webster and Cherry, 1978) on top of Saevar Howe have shown several superimposed phases of the pre-Norse and Norse periods with an abundance of animal remains, these have yet to be analysed. The following observations are based on preliminary analysis of the material from excavations by Durham University on the Brough of Birsay up to and including 1976, and from sites excavated on the mainland side of the bay, related to the published material from Dr Ritchie's site of Buckquoy. Only where it can be clearly established on archaeological grounds, will distinctions between the pre-Norse and Norse periods be made.

An unfortunate aspect of some of the excavations on the Brough is the very poor state of preservation of the bone and shell. Unlike the mainland edge of the bay, the Brough has no shell-sand deposits and archaeological layers have been buried in peat or an acid turf. The soils have developed on boulder clay, drainage is poor, and the soil acid in places periodically saturated. However excavations in Room 5 in the complex of standing structures on the eastern cliff area of the Brough, were in an area with good preservation, and much of the preliminary analysis of this material is complete. The site has at least four structural phases and wood charcoal of willow or poplar from the third phase has been given a radiocarbon determination (uncalibrated) of 955 ± 60 ad (GU-1193). Apart from evidence for minor industrial activities on the site including bronze working and iron working, the main interest of the site lies in environmental data. The animal bones and shell permit some comment on the subsistence economy of the site. To the west and south-west of the cemetery are apparently scattered rectangular and sub-rectangular buildings (see Fig. 3). Site L, excavated in 1975, was a dwelling with at least two, if not three, major phases (as represented by the replacement east wall) surrounded by an open area traversed by gulleys and drains and some small archaeological features, such as a pit with remains of iron-smithing, to the east of it. Site N, to the east, appeared in 1974 to have no surviving internal features to suggest a function other than that of a store-house. However, subsequent excavation showed that it overlay earlier features, possibly even a building, and certainly pits associated with minor industrial activity and a gulley. Three radiocarbon determinations on these layers gave uncalibrated dates of 815 ± 55 a.d. (GU-1192 & 1195) and 750 ± 85 a.d. (GU-1194). Finds in and around these buildings were few in number, and no middens were found in association with them. Sites E and S, on the otherhand, have, in three seasons of excavation, displayed a somewhat baffling structural history, with several phases and sub-phases of clearance and re-building. On Site E, the latest levels of outhouses associated with a small building have a *terminus post quem* of the first part of the Eleventh century on the basis of a coin found below one of the walls, and appear to lie on clearance layers with earlier material such as a bronze mount with a fine interlace pattern on it of pre-Norse Insular type. These clearance layers include an amount of mammal bone, but its survival and condition in a clay matrix is poor, and dry and wet-sieving has sometimes proved either negative or counter-productive (because of the total collapse of the bone). Beneath these layers are remains of earlier buildings, but examination of these is not yet complete. Site S is particularly complex, and rich in both structural features and artefacts. The most characteristic Viking period artefacts have been recovered from this area, including steatite bowl fragments, net-sinkers, and bone-comb fragments. Environmental data have been collected, but often the problems are the same as those for Site E.[3]

The basic subsoil of the landward side of Birsay Bay is largely sand rather than clay, and therefore more conducive to environmental processing. Some fine mesh sieving and flotation has been carried out on material from Areas 1 and 2, beside the Brough Road (see Fig. 4). These areas were excavated because of depletion of the sand-cliffs by winter storms, and the obvious existence of extensive midden material. This midden clearly contained Viking period artefacts and indeed had one or two graves with such artefacts dug into it. It also overlay the remains of both flagged areas and walls of buildings, largely under the adjacent road and field. A cist grave with the partial remains of two skeletons was salvaged in 1976, and probably stratigraphically predates this midden. It also probably is contemporary with a second cist-grave, again without grave-goods, found below a mound of sand under a stone cairn, with carefully constructed kerb. This cairn certainly pre-dates the midden, and so is tentatively thought of as pre-Norse, pending radiocarbon determinations on the bone.

Also possibly pre-Norse were buildings located approximately 100 m away and also being eroded. In Areas 3, 4 and 5 a complete figure-of-eight shaped building was exposed, probably with a corbelled roof (from the position of the rubble collapse, entirely internal), and adjacent circular end of a building, now lost to the elements. The structural points of comparison of this latter building to the end of Dr Ritchie's 'Later Pictish' building (a mere 90 m away) suggests a similar date, but again absolute dates are awaited. In this case the carbonised grain from pits cut into the natural clay below a hearth in one of the rooms of the figure-of-eight house should provide a fixed point.

In the village of Birsay itself, the site of Beachview is being excavated, partially because of building proposals and partially because of winter flooding by the Burn of Boardhouse. The Burnside sites include a sequence beginning with a structure represented by two walls dug into and bedded onto the natural clay and flagstone. Sand layers above seal this structure from the destroyed remains of another, itself sealed by a large band of sand below midden-like deposits, partially excavated in 1979. Further along the Burn-side to the east are exposed further rubble spreads, burnt clay layers and midden material. To the south-west, however, on the 'Studio' site a major rectangular building is now being revealed. Preliminary indications are that it may be c. 5 m wide and up to 15 or 20 m long. Above it the partially destroyed remains of a small circular building with a central slab-lined hearth are interesting, particularly because they contained many carbonised grains; the midden material was discovered spreading down into the rectangular building, below the later rubble collapse. Artefacts in this midden layer suggest a late Norse (Viking/Medieval) date on analogy with material from the site of Freswick, Caithness (Curle, 1939; Childe, 1943)—indeed Freshwick-type grass-tempered pottery has been found, and parts of combs similar to those found at Freshwick. A radiocarbon determination of 1010 ± 55 a.d. (GU-1191) on carbonised grain and charcoal has been received. The walls of the building are standing c. 1 m high, with seven or eight courses of stone, but the floor-level is not yet reached. Excavation will continue on the site.

Environmental Evidence

(a) Vertebrates

Turning to the environmental data as such from the various sites, of these later periods, around the bay, Seller's analysis of the animal bone from Room 5 on the Brough (unpublished) has indicated a change in the domestic husbandry of the site from the earliest pre-Norse phase to phase 4 in the Norse period. This involves an increase in the slaughter of young animals in the latest phase (65% of the cattle were older than 3½ years in phase 4) and a much lower percentage of adult animals in comparison with phase 1 (80% of cattle over 3½ years in phase 1). This is somewhat in contradiction to the site of Buckquoy spanning a similar period of time (Noddle, 1979). Analysis of that material indicated that in general a third of the animals had died in their first year and two thirds by the fourth year, with a higher proportion of cattle and sheep reaching maturity in the later periods. A further difference between the two collections is that sheep are the most numerous bones in all phases at Room 5 (although not the most important) whereas cattle are consistently the most frequent fragments at Buckquoy. A more detailed analysis of these differences must await the completion of the current analyses on the sites adjacent to both of these. However, the latter differences may in part be accounted for by the different collection techniques: the Room 5 site was wet-sieved on the lines of those put forward by the "Cambridge School" using a mesh size of 5 mm; whereas no sieving was carried out at Buckquoy; (Ritchie, 1979, 191). Besides domestic animals, remains of wild species such as deer occur occasionally in deposits of this period in the bay (Noddle, 1979; Rackham, unpubl.). Although it is often in the form of antler, a material quite likely to have been imported for the manufacture of articles (a number of which have been found on excavations at Birsay), other bones are found.

Marine mammals tend to occur more abundantly than the terrestrial wild food species. Whalebone has been found at Buckquoy (Noddle, 1979). and excavations by the project in 1976 on the Brough, and in 1978/9 at the Beachview Studio site produced further finds from the Norse period. Unfortunately many of these pieces are fragments so that species determination is not possible, or it is made more difficult by the fact that a number are 'worked'. Noddle (1979) has suggested that such remains come from washed up specimens, and the example of a dolphin washed up in the bay during the excavation season in 1979 illustrates the point. The pilot whale, *Globiocephala melaena*, known locally as the Blackfish or *ca'in* whale is the common whale off the Northern Isles, and Fenton (1978) notes that schools of these whales were driven ashore by the inhabitants of the islands. The importance of the animals as a source of oil (worked bones of whale have been found, coincidentally, associated with a steatite lamp in one context on Site S on the Brough) is noted as early as the sixteenth century, and presumably local inhabitants would have taken advantage of sporadic schools by driving them ashore even in the Norse period. Ohthere's report to King Alfred of how with five companions he had killed 60 whales in two days (Ross, 1940) is reminiscent of such an occasion although the sizes he gives for these animals are, as far as we understand his units, far to large for the pilot whale. The importance of whales in prehistoric Europe has been discussed by Clark (1947) and in

Prehistoric Europe (Clark, 1952) he notes "Olaus Magnus maintained that the proceeds of a single whale might fill between 250 and 300 waggons and yield meat for salting, blubber for lighting and heating, small bones for fuel, large ones for house-building and hide sufficient to clothe forty men". Animals slaughtered and butchered on the beach are likely to leave only those bones saved for a purpose on a habitation site, and therefore cannot reflect the importance of the catch. The large sperm whale, Physeter macrocephalus, was identified at Jarlshof (Platt, 1956) and the salting of such whalemeat, noted by Evans (1957) as being practised in Ireland, would be a considerable source of food as Olaus Magnus maintained.

The grey seal, Halichoerus grypus, and the common seal, Phoca vitulina, are our most common Pinnipedia and both species currently breed in Orkney. Both species have been identified from Norse settlements in the Northern Isle and at Jarlshof the grey seal, at least, was as frequent an occurrence as the bones of domestic animals in some layers of ninth to eleventh century date (Platt, 1956). Although species identification has not been made, seal bones have been found on all the sites so far studied in Birsay Bay. However they do not suggest any greater importance that those recovered from the excavations at Buckquoy, where they were an infrequent occurrence in most phases (Noddle, 1979). Fenton (1978) notes the exploitation of these animals for fat, meat and skins, and Evans (1957) has noted the skins as being used in Ireland for making the shoes rock climbers used, presumably when climbing the cliffs for wildfowl eggs and young.

The otter was an incidental find in all phases on the Room 5 site, and in one deposit a partial skeleton is probably present since forty-nine bones were identified (Seller, unpubl.). The otter is semi-marine in the northern parts of its range in the British Isles, being seen frequently in coastal waters. They have been sufficiently common in the Northern Isles in the past for substantial stone traps, otter-hooses, to be built across otter's tracks. The skin's, a valued commodity, were being exported from Orkney in the seventeenth century (Fenton, 1978) and it is possible that such pelts were collected and traded in pre-Norse and Norse times.

An important aspect of the coastal exploitation, for which we as yet have little specific data, is wildfowl. The bones of gannet and other wild bird species are abundant on the excavations being carried out by the project but the analysis of these is still in its early stages. The species identified from the prehistoric site on the Point of Buckquoy mentioned above included birds of marsh, meadow and freshwater as well as coastal habitat. Waders are the most abundant and include golden plover, sandpiper and redshank. The published species from Buckquoy (Bramwell, 1979) show that a wide range of wild birds were exploited from a variety of habitats, all of which are local to Birsay. There is some indication that this exploitation is more marked and varied in the later Norse phases but this may very well be a feature of sample size. A number of species present today on the cliffs and rocks immediately north and south of the bay were found, including razorbill and guillemot, both present in their thousands on the cliffs, but also puffin, fulmar, cormorant, shag and gannet. The last species was the most frequent find in all phases of the site. Wildfowling has a documented history in Orkney since the seventeenth century (Fenton, 1978) and its antiquity is indicated by the finds from Buckquoy and

Cutting 6. The collection of eggs and juvenile birds from inaccessible cliffs by letting oneself down or being let down, by a rope is traditional and on some offshore islands such as Foula in Shetland, were a major or the major support for the inhabitants. The cliffs around the bay at Birsay are ideal for such a harvest and the food, oil and feather were no doubt important resources in the past.

Probably the most important marine resource of the sites of the Norse period in the bay and elsewhere in the northern islands and coastal lands of northern Scotland is fish. We have noted above the exploitation of this resource in the prehistoric period in the bay when cod, saithe, mackerel, rockling and wrasse were among the fish caught most probably from inshore or onshore fishing. This resource has been exploited in Orkney since the earliest times of known human occupation (Clarke, 1976; Wheeler, 1979a). Both Seller's unpublished report of the Room 5 group and Wheeler's results from Buckquoy (Wheeler, 1979b) illustrate the exploitation of these animals during pre-Norse and Norse times. Seller's work has suggested a major change in this aspect of the economy during the period of occupation of Room 5 and the figures given by Wheeler seem to reflect this change. In phase 4 at Room 5, cod constitute over 25% of the identified bones, an increase of over 20% on the early phases of occupation, and Seller has suggested that the marine resource during this period was at least as important as the sheep flock. Other species included mackerel and ray. At Buckquoy, despite the absence of sieving on the site, Wheeler (1979b) has identified a wider range of species, conger eel, cod, saithe, haddock, ling, hake, wrasse, seabream, mackerel and gurnard. There is some evidence from the relative abundance of the remains at Buckquoy that fish became a more significant element of the excavated fauna in the last two Norse phases of occupation at the site. The difference in species and size of the fishes caught has suggested to Wheeler that improved fishing techniques prosecuted in deeper water, using many hooked lines, would account for the observed differences. This is very much a traditional method, that of haaf fishing, in Orkney (Fenton, 1978). An improved maritime technology of the Norsemen over that of the native populations of the Northern Isles may well account for such a change, and is also suggested by a number of other sites in the north. The remains from the early ninth century midden of the Norse settlement at Jarlshof (Platt, 1956) included cod, saithe and ling, mainly fish of very large dimensions with skulls alone measuring over 0.3 m; although both cod and ling are identified from the earlier levels at the site, the published report suggests that these specimens are not of notable size. This change has also been suggested on artefactual grounds at both Underhoull, Unst (Small, 1966) and Jarlshof (Hamilton, 1956). Work in recent seasons at the late-Norse settlements of Sandwick, Shetland (Bigelow and McGovern, 1979) and Freswick, Caithness (Rackham, unpublished report) has also produced a very high percentage of fish bones in midden material, in places forming almost continuous layers, in conjunction with winkle and limpet shells at the former site. The analysis of the preliminary samples from Freswick and the finds from the 1979 season at Beachview, Birsay (Colley, pers. comm.) both indicate cod and ling bones from large fishes: many at Beachview were over 1 m long and all at Freswick. The evidence from Room 5 and Buckquoy in the bay suggests that this change is not coincident with the onset of Norse occupation, but rather with later phases of the recognised occupation.

(b) Invertebrates

The seashore and littoral zones have traditionally been areas of resource in Orkney. A considerable number of the subsistence activities carried out both in the past and more recently have relied on the produce of these elements of the coastal environment. The shellfish and crustaceans frequent on both sandy and rocky shores along the coasts of these islands have been a continual source of food since Neolithic times (Clarke, 1976). It is probably wrong to assume that all the shellfish, or even the majority, from the pre-Norse and Norse deposits in the bay were utilised for food. Evans (Evans and Spencer 1979) has shown that the total flesh from the 100,068 mollusc shells from Buckquoy, the majority of which were limpets and winkles, would have been only sufficient to support one man for ten days. Even as a sporadic addition to the diet, such a return of nutrition for labour of collection is not encouraging. However, although sometimes used as a food by the poor and in times of hardship (Fenton, 1978), all manner of shellfish, and particularly the limpet and periwinkle, which are the most common ihhabitants of the rocky shores of Birsay today, have been traditionally used as bait. The limpet, parboiled, formed the most common bait for craig-fishing (from the rocks) (Fenton, 1978), being chewed to a pulp and spat out to form an oily film on the water that attracted the fish. The very high number of winkles and limpets collected at the excavations at Buckquoy from Norse phases IV and V in comparison with the number from earlier phases (Evans and Spencer, 1979) may be related to the suggested increase in the importance of fishing and the change in techniques. The limpets are used as bait to catch the saithe and haddock which form the traditional bait for haaf (ocean) fishing, with deep lines and many hooks, for cod and ling (Fenton, 1978). A similar abundance of limpet and winkle have been identified by Woodward (unpublished data) from the molluscs from Room 5, and, though still to be analysed, molluscs are more abundant than bones from almost all the excavations since 1975 that have been carried out in the bay. Some shellfish we assume must have been used for food, particularly the oyster, cockle, scallop and probably mussel and horse-mussel: some of these may have been dredged for, and all occur on these sites in Birsay bay. The area of collection appears to have included the sub-littoral as well as the littoral zone.

(c) Seaweed

A further aspect of the mollusc remains little considered until recently (Evans, 1979; Evans and Spencer, 1979; Bell, this volume), is the possible light they may shed on the exploitation of seaweed. A number of the smaller species of mollusc (unlikely to have been collected for food or bait) feed on, or are commonly found browsing over, the fronds of seaweed, while others have a tendency to attach themselves to the holdfasts of the larger seaweeds such as <u>Laminaria</u> species (Step, 1945; Fretter and Graham, 1962). Of these species a number were found in Room 5 (Woodward, unpublished list), and four species occurred at Buckquoy (Evans and Spencer, 1979); they include <u>Gibbula cineraria</u>, <u>Calliostoma zizphinum</u>, <u>Patina pellucida</u>, <u>Chlamys distorta</u>, <u>Hiatella arctica</u>, <u>Littorina neritoides</u>, <u>L. saxatilis</u> and <u>L. littoralis</u>, which is usually found on <u>Ascophyllum nodosum</u>, the knotted wrack, or <u>Fucus vesiculosus</u>, the bladder wrack. <u>F. vesiculosus</u> has been identified from a probable Norse context in Site S on the Brough of Birsay. The context contained a considerable volume of fragments of carbonised seaweed, mostly of the flattened fronds with no diagnostic features, but swollen, granular frond fragments, frond tips and an air bladder were identified as bladder wrack and all the material could be referred to this species (see Plate 1). A few frond fragments also referrable to the bladder wrack were found in the probable Norse midden of Area 2 beside the Brough Road. Seaweed has also been recognised among carbonised material in pits in a supposed cooking area at The Biggings, Papa Stour, Shetland (Crawford, 1979a). Seaweed has been, and sometimes still is, put to a variety of uses in Orkney, particularly manuring but also, fuel and food. Mucilaginous seaweed has been used for caulking boats, its hygroscopic properties preventing the seams drying out, even when the boat is out of the water (Dimbleby, 1978). The bladder wrack identified from the Brough is an edible species and certain species are still gathered for human consumption in Ireland, i.e. *Chondrus crispus*, *Porphyra* sp. and *Rhodymenia* sp. (Evans, 1957). In Orkney it was also used sometimes as cattle fodder: *Alaria esculenta* was parboiled before being fed to the animals, and there is the well known situation on North Ronaldsay where the sheep graze on seaweed for the whole year (Fenton, 1978).

A more recent use of seaweed, the burning of kelp on a large scale to produce a residue rich in potassium and iodine salts, is also evident in the archaeological record around the bay. The ash of wracks such as *Fucus* and *Ascophyllum* was used in the past for the production of glass and iodine (Fenton, 1978). In the field south of the Beachview Studio Site, two kelp pits are visible, if a little overgrown by turf, and closely resemble the kiln excavated by Fenton (1978) on North Ronaldsay. One or two similar depressions, but much overgrown, that occur nearby on Birsay Links as well as some beside the road to the Brough are almost certainly similar features. Two features interpreted as kelp kilns were observed in the dunes north of Saevar Howe where they had been exposed by quarrying. The better preserved of these consisted of a number of stones and burnt material lying below a 0.1 m thick turf layer and a lower 0.5 m depth of wind-blown shell-sand. The burnt material was spread up to six upright burnt stones which formed the corner of a disturbed feature. Although smaller than the North Ronaldsay example, measuring internally 0.87 m by 0.76 m, with upright stones varying from 0.01 m to 0.08 m thick, the structure is similar in form. Samples taken from the spread of carbonised material have still be to analysed. As yet none of these pits can be dated, but the industry was mainly an 18th and 19th century one with some records from this century also (Fenton, 1978).

(d) Agriculture

An aspect of the coastal environment of importance in the agricultural economy of the bay, and concievably also in the pre-Norse and Norse periods, is the quarrying and spreading of shell-sand on the land as a fertiliser. The soils at Birsay are by tradition the most fertile on the island (Ritchie, 1975) and no doubt wind-blown and carted shell-sand have been important elements, along with the ready supply of seaweed as manure, in the establishment of this tradition. The antiquity of the tradition, which incidentally has been causing severe erosion in the bay in recent years (Mather *et al.* 1974), is unknown, and Fenton makes no mention of it. The manuring of fields would have assisted the growing of cereal crops if carried out during the pre-Norse and Norse occupation of the bay.

Little grain has so far been recovered from the samples taken from the Brough sites, but the middens and floor levels of the two major areas excavated by the side of the Brough road (Areas 1 and 2, and 3 and 4) as well as the Beachview site, have yielded a considerable volume of grain and a few other carbonised seeds. A hearth in the figure-of-eight house in Area 3 contained over three thousand grains of barley: a few exceptionally well preserved grains and a rachis enabled identification as Hordeum vulgare L. (bere), the hulled 6-row variety which was the primary crop in Orkney until at least the eighteenth century. It is still grown, only in Orkney, to a very limited extent, but has elsewhere been superceded by 2-row varieties (H. distichon L.) Grain from the midden deposits in Areas 1 and 2, though less well preserved, is also referable to bere. A small amount of oats was also present in Area 3 in the figure-of-eight house. Although grain alone is not sufficient for specific identification, four flower bases were identified as Avena fatua (wild oat). Generally a common weed of cereal crops, it is also an introduced species with recorded occurrences dating from the Iron Age, and it is thought to have been introduced with grain seed (Jessen and Helbaek, 1944). The supposed Norse middens of Areas 1 and 2 also contained a small proportion of oats: the only flower base recovered was identified as Avena sativa, the common cultivated oat, thought to be derived from A. fatua, also a hexaploid. The grain in these middens has a slightly larger size range also pointing to A. sativa, though A. fatua is probably still present. A. sativa is also recorded from the Iron Age onwards and probably originally grew as a weed in cultivated fields, with its potential as a crop in its own right being realised in the more northerly latitudes. As Ritchie (1979) pointed out, according to the Statistical Accounts of 1795, bere and oats have been the most important crops in Birsay since the medieval period. Very few seeds other than grain have been recovered and they add little to the environmental picture. Carbonised seeds of sedges, docks, grasses and goosefoot have been found sporadically throughout the periods represented, and are probably from the local fields, dunes and beaches. One sample which contains much grain, yielded several seeds of the corn spurrey, a frequent weed of crops on sandy soil, and which was certainly used in its own right with the seeds of other weeds for making bread (Fenton, 1978).

(e) Other Resources

Among the other carbonised remains obtained from the pre-Norse and Norse excavations on the Brough and beside the Brough road, are considerable amounts of charcoal, often alongside burnt peat. Willow or poplar, birch, hazel, pine and oak have been identified. Although there is some doubt about the status of pine, oak is almost certainly not native to the islands. The small amount of these timbers could have been imported or collected as driftwood, but the large amount of branches and twigs of the other species suggests the use of local material, and may be evidence for the presence of some scrub woodland in Birsay at least up to the Norse period. Although driftwood may have been a useful resource in the past, with quite large tree trunks washed up, the possibility of imported timber in this period, from Scandinavia or Scotland, cannot be forgotten. The find from the Biggings, Papa Stour, of a planked floor of pine in a late-Norse building has already led to the suggestion that the timber came from Norway (Crawford, 1979b), and the roof supports that are suggested by the structural remains at Buckquoy may perhaps have necessitated import from Scotland (Ritchie, 1975).

A further, almost too obvious, resource of the foreshore and beach is stone. Fire fractured beach pebbles are a common find, occurring especially at Beachview in Birsay but also at other sites of the Norse and late-Norse period in the bay and also at Freswick. The use of such stones at Papa Stour for 'parching' grain has been suggested by the finding of large stone hearth slabs upon which grain might have been laid to dry (Crawford, 1979b), and Fenton (1978) has suggested this use as an alternative to the more traditional view of the stones as 'pot boilers'. Their use as a source of heat in steam-cooking in sealed pits has been suggssted at Papa Stour (Crawford, 1979a). The foreshore in Birsay bay, of gently inclined flagstones, is also an ideal source of building stone (Ritchie, 1975) and would form the closest and most easily worked quarry. These rocks are readily split with hammers and wedges (Fenton, 1978) and there is no reason to suppose that this simple quarrying was not within the technological capabilities of the pre-Norse and Norsemen.

Conclusions

The emphasis in this paper, as in the conference, has been upon marine and coastal resources. However, it would be a misrepresentation to suggest that the economy during the pre-Norse and Norse period was either based on, or largely dependent upon, these resources. The importance of agriculture and animal husbandry, if not discussed at length above, should not be underestimated: the parallel evidence from Jarlshof and Underhoull, in Shetland (Hamilton, 1956; Small, 1966) suggests it was basic. Equally, although from the concentration of known Norse settlements in Faroe and Shetland along the coast, it has been concluded (Small, 1968) that they appear to have been arranged with respect to the need for access to the sea as well as land for crops and animals, it is difficult and dubious to draw a similar conclusion for Orkney. It is an observable fact that the landscape of the Orkney Islands is very different from that of the other Atlantic Island groups, and much more archaeological and environmental investigation must take place before firm conclusions about patterns of settlement in these islands can be made.

This paper must be seen as a preliminary review of work in progress, and an indication of the future possibilities from the work of this project. Other aspects, such as those of the relationship of artefactual and environmental material, and the possibilities of evidence for trading, will need to be considered in future work. Much more sorting and analysis of material recovered so far, as well as that from future excavations, needs to be done. In time for instance the steatite lamp and the line or net-sinkers for fishing from probable Norse contexts on the Brough will have a clearer context in a pattern of life, which at present we can simply glimpse or guess at.

Also, as time goes on, recovery methods become more efficient. The molluscan and animal bone material from Room 5 on the Brough and Dr Ritchie's excavation at Buckquoy have yielded valuable results, but more systematic and adequate recovery methods than were attempted on those sites are necessary for other data. The evidence of fish, bird, cereals and wild plants is, as indicated above, obviously important for the light they throw upon the utilisation and exploitation of the environment at Birsay; and these fields of evidence depend very much upon adequate recovery for any reasoned assessment of their importance.

This year on the Beachview Studio Site a small team excavated the midden within the longhouse of Norse date. Recovery was effected by the total wet-sieving, through 2 mm and 0.895 mm mesh sieves, with the aid of water pumped from the burn. Some attempt was made to measure the volume of the material sieved and the fine fraction (greater than 0.895 mm) was floated and stored separately. The recovery of large numbers of carbonised cereal grains, largely unnoticed during excavation, and the efficient recovery of fish bones, was immediately appreciated, and it is to be hoped that these recovery methods will enable a much more detailed analysis, of considerably larger samples, to be made on the material from the most recent excavations and those of the future within the project.

Finally, it should be noted that in a project of this nature, the wider context of the bay has to be considered alongside the specific detailed evidence of particular excavations. The potential for enlarging understanding of the nature, range and mechanisms of settlement from small-scale sampling has already been demonstrated by the results from Cutting 6. The processing of further midden samples from other exposed, unexcavated, sites, is being undertaken, and may confidently be expected to supplement and expand existing information, and indeed give an indication of the archaeological potential of these sites. Several of the sites mentioned above were exposed by winter storms in 1976, and further archaeological features were noted following widespread destruction of sandcliffs by a hurricane in 1978. All sites excavated, sampled, or simply noted, to date are included on the accompanying map of the Bay (Fig. 5), and it cannot be regarded as in anyway a complete representation of the archaeology of Birsay: further work has to be done inland, and in the southern part of the bay. It cannot be claimed that the situation is unique, but perhaps the concentration of resources in this area is unusual, and the potential great. As the project develops, it is also hoped to compare the material from this area of Orkney with the equally rich, and equally threatened, material in the area of Freswick, Caithness, and the parallel project at Sandwick, Unst, Shetland (Bigelow and McGovern, 1978). Such concentrated archaeological and environmental projects seem to offer an exciting prospect for enlarging our understanding of the interaction of man and the environment in the past.

NOTES

1. The excavation in 1973, by Mrs Curle and assisted by J. R. Hunter, of Room 5 in the Eastern Cliff Settlement, was completed by J. R. Hunter and C. D. Morris in 1974. From 1974 work by Durham University under the direction of C. D. Morris has taken place on an area to the west of the churchyard (Sites L, N, E and S) and on the Southern Cliff Area. Since 1975 Dr Hunter, now of Bradford University, has undertaken excavations on the Northern Cliff Area (Sites, F, P and R) and the area immediately south of the Guardianship Area. All work has been financed by the Ancient Monuments Inspectorate of the Scottish Development Department (formerly D.o.E), and Durham and Bradford Universities.

2. These excavations by Durham University, directed by C. D. Morris, assisted by N. Emery and N. F. Pearson, were financed by the Ancient Monuments Branch of the S.D.D. and Durham University.

3. Since the observations on the environmental data are confined to material from Room 5 and the area to the west of the churchyard, no description is here given of the other areas on the Brough under excavation either by C. D. Morris or Dr. J. R. Hunter.

Acknowledgements

A considerable number of people are contributing to the work of the project and we should like to thank those that have permitted their data or results to be used in this paper. These include Dr S. Limbrey, A. K. G. Jones, N. Emery and Dr F. R. Woodward, and in particular we are grateful to Dr T. J. Seller for permission to use his report prior to publication. The material from the following unpublished reports has also been utilised:

Alvey, B. A. P., Bramwell, D., Donaldson, A. M., Rackham, D. J. and Wheeler, A. 'Preliminary Study of a prehistoric site at Birsay, Orkney; in advance of excavation'.
Rackham, D. J. 'Freswick: a recommendation for excavation'.

We would also like to thank N. Emery for the drawings and the Photographic Dept., Department of Archaeology, University of Durham for the plates. Thanks are due to Dr D. J. Breeze and P. J. Ashmore, Inspectors of Ancient Monuments, S.D.D., for their help and support.

(All illustrations are Crown Copyright).

BIBLIOGRAPHY

Bell, M., (this volume).

Bigelow, G. F. and McGovern, T. H., 1978. Preliminary report of the 1978 Excavations at Sandwick, Unst, Shetland Islands. Norse Shetland Economic Research Project. Privately circulated.

Bramwell, D., 1979. Appendix 2. Bird and vole bones from Buckquoy, Orkney in Ritchie, A. (1979) see below.

Childe, V. G., 1943. Another late Viking house at Freswick, Caithness, Proc. Soc. Antiq. Scot., LXXVII: 5-17.

Clark, J. G. D., 1947. Whales as an economic factor in Prehistoric Europe Antiquity, 1947: 84-104.

Clark, J. G. D., 1952. Prehistoric Europe, The economic basis, Methuen, London.

Clarke, D. V., 1976. The Neolithic Village at Skara Brae, Orkney: 1972-73 Excavations: an interim report. Edinburgh.

Crawford, B., 1979a. A progress report on excavations at 'Da biggins', Papa Stour, Shetland, 1978. Northern Studies, 3: 37-41.

Crawford, B., 1979b. Third Interim Report of the Excavations at The Biggings, Papa Stour, Shetland, 1979. Privately circulated.

Curle, A. O., 1939. A Viking settlement at Freswick, Caithness. Proc. Soc. Antiq. Scot., LXXIII: 71-110.

Davidson, D. A., Jones, B. L. and Renfrew, C., 1976. Palaeoenvironmental reconstruction and evaluation: a case study from Orkney. Transactions of the Institute of British Geographers, 1: 346-361.

Dimbleby, G., 1978. Plants and Archaeology. Baker, London. 2nd ed.

Evans, E. E., 1957. Irish Folk Ways. RKP.

Evans, J. G., 1979. The palaeoenvironment of coastal blown-sand deposits in western and northern Britain. Early man in the Scottish Landscape, Scottish Archaeological Forum, 9, 16-26.

Evans, J. G. and Spencer, P., 1979. Appendix 4. The mollusca and environment, Buckquoy, Orkney. In Ritchie, (1979).

Fenton, A., 1978. The Northern Isles. Donald, Edinburgh.

Fretter, V. and Graham, H., 1962. British Prosobrach Molluscs, Ray Society, London.

Hamilton, J. R. C., 1956. Excavations at Jarlshof, Shetland, H.M.S.O. Edinburgh.

Jessen, K. and Helbaek, H., 1944. Cereals in Great Britain and Ireland in prehistoric and early historic time. K. Danske Vidensk. Selsk., 3: 1-65.

Keatinge, T. H. and Dickson, J. H., 1979. Mid-Flandrian changes in vegetation on Mainland Orkney. New Phytol., 82: 585-612.

Lamb, R. G., 1975. The Cathedral of Christchurch and the monastery of Birsay, Orkney. Proc. Soc. Antiq. Scot, 105: 200-205.

Marwick, H., 1970. The place-names of Birsay. Aberdeen.

Mather, A. S., Ritchie, W., and Smith, J. S., 1974. Beaches of Orkney. Report for Countrside Commission for Scotland, pp. 168, Department of Geography, University of Aberdeen.

Moar, N. T., 1969. Two pollen diagrams from the Mainland, Orkney Islands. New Phytol., 68: 201-208.

Mykura, W., 1975. The Geological Basis of the Orkney Environment. in The Natural Environment of Orkney, ed. R. Goodier, Nature Conservancy Council, Edinburgh.

Noddle, B., 1979. Appendix 1. The animal bones from Buckquoy, Orkney. in Ritchie, A. (1979).

Olaus, Magnus 1555. Historia de gentibus septentrionalibus. Rome.

Palsson, H. and Edwards, P. (trans.), 1978. Orkneyinga Saga. The History of the Earls of Orkney. Edinburgh.

Platt, M. I., 1956. Appendix IV. The Animal Bones. in Hamilton, J. R. C. (1956).

Prentice, H. C. and Prentice, I. C., 1975. The Hill vegetation of North Hoy, Orkney. New Phytol., 75: 313-367.

Renfrew, C., 1979. Investigations in Orkney. Reports of the Research Committee of the Society of Antiquaries of London, No. XXXVIII, pp. 234.

Ritchie, A., 1975. Pict and Norseman in Northern Scotland. Scottish Archaeological Forum, 6: 23-36.

Ritchie, A., 1979. Excavation of Pictish and Viking-age farmsteads at Buckquoy, Orkney. Proc. Soc. Antiq. Scot., 108: 174-227.

Ross, A. S. C., 1940. The Terfinnas and Beormas of Ohthere. Leeds.

Small, A., 1966. Excavations at Underhoull, Unst, Shetland. Proc. Soc. Antiq. Scot., XCVIII: 225-248.

Small, A., 1968. The distribution of settlement in Shetland and Faroe in Viking times. Saga Book, XVII: 145-55.

Step, E., 1945. Shell Life, Warne.

Webster, L. E. and Cherry, J., 1978. Medieval Britain in 1977. Med. Arch., XXII; 152.

Wheeler, A., 1979a. Appendix G: The fish bones. in Renfrew, C., (1979). Ch. IX, 144-149.

Wheeler, A., 1979b. Appendix 3: The fish bones from Buckquoy, Orkney, in Ritchie, A. (1979).

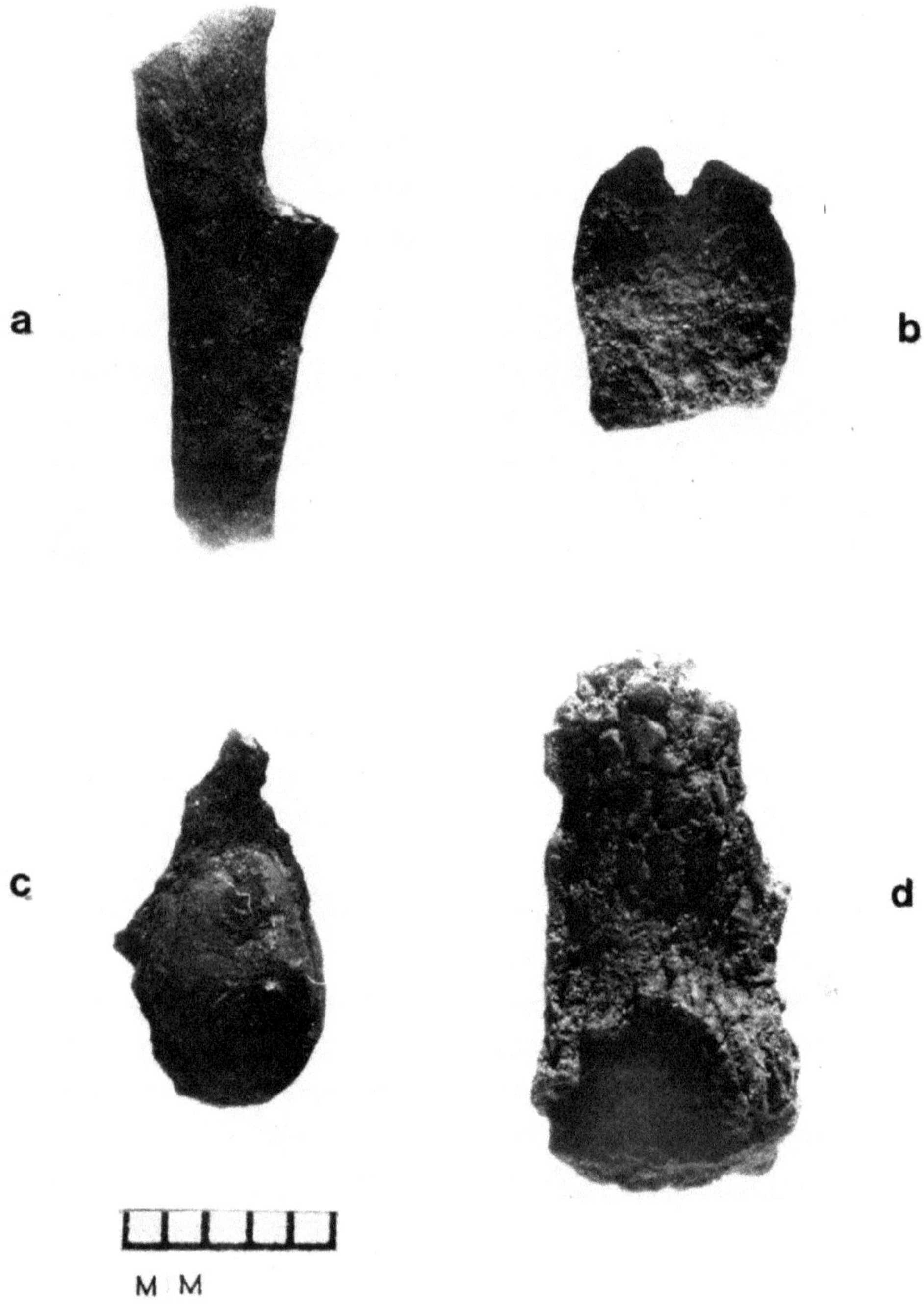

Plate I Carbonised fragments of bladder wrack, <u>Fucus vesiculosus</u>, from a probable Norse context on the Brough of Birsay.

a) Node b) Frond tip c) Bladder d) Swollen, granular fruiting body

THE PALAEOECOLOGY OF ESTUARINE DEPOSITS ASSOCIATED WITH ARCHAEOLOGICAL SITES WITH PARTICULAR REFERENCE TO THE CITY OF LONDON

By Peter D. A. Boyd

SYNOPSIS

The range of archaeological sites situated on estuaries is large. Estuaries have formed a route for traders and invaders encouraging the development of trading settlements and defensive installations. The abundance of fish, shellfish and bird life supported by estuaries has encouraged the development of communities based on the cropping of these natural products. Salt-making has also been an important activity on the banks of estuaries as shown by the occurrence of numerous 'Red Hills'. Many of the world's major cities have therefore developed on estuaries.

The City of London was founded on the north bank of the River Thames in the first century A.D. at a point which was probably close to but above the limit of tidal influence at that time. The mean sea-level has risen since then and continues to rise. The erection of successive waterfronts combining the need for higher revetments with reclamation has narrowed the channel and exasperated the problem of flooding.

The Department of Urban Archaeology of the Museum of London has pioneered work on waterfront archaeology since its formation in 1973 and attempts have been made to estimate the 'sea-level' at various times using the height of excavated waterfront structures relative to ordnance datum. However, little work was done on the biological remains from estuarine deposits associated with these waterfronts except those remains of 'cultural' interest dumped there. An international conference on the Archaeology of Waterfronts in North-European Towns held in London in 1979 showed this to be so elsewhere as well.

Therefore the author was asked to carry out a preliminary examination of these estuarine deposits to assess their potential as sources of environmental information (biological and sedimentological).

A large variety of biological remains is recoverable if suitable techniques are adopted. Most remains are organic, calcareous or siliceous but 'trace-fossils' are also found. The remains recovered from London waterlogged deposits include organic-walled microplankton, diatoms, charophytes, seeds, pollen, testate rhizopods, foraminifera, sponge spicules, bryozoan statoblasts, parasitic worm eggs, cladoceran ephippia, ostracods and molluscs. Many of the organisms from which these remains are derived are sensitive to small changes in salinity, pollution and other environmental parameters.

Salinity, pollution, tidal amplitude, current velocity and substrate were the main factors which, influenced by long-term environmental changes, civil engineering projects and waste-disposal practices, determined the composition of the estuarine biocoenoses. However, post-mortem transport of subfossil remains, reworking of Tertiary and Mesozoic microfossils and post-depositional processes have also affected the composition of the assemblages of biological remains preserved.

The Thames was a major source of water for drinking and other domestic uses until medieval times when a combination of high salinities and pollution encouraged the City authorities to pipe fresh water in from wells and springs on the outskirts of London. Salinity changes and pollution also affected other activities based on the Thames and its tributaries and it became a serious health hazard. Documentary records give part of the story but it is only through scientific studies of the sedimentary deposits that it will be possible to examine the ways in which the estuary has affected human activities during its history and the effects which Man has had on the estuary and its flora and fauna.

The contents of this paper are being published in a much extended form elsewhere.

ADAPTATION TO A DESERT COAST: SUBSISTENCE CHANGES THROUGH TIME IN COASTAL PERU

Barbara Pickersgill and Richard T. Smith

INTRODUCTION

The Incas referred to their empire as Tahuantinsuyu, land of the four quarters. Although contrasted environments of the Andes (sierra), equatorial rain forest (montana and selva) and desert coast (costa) come immediately to mind, the four quarters were actually administrative provinces equivalent to the four sectors of the compass. In Inca times the whole area functioned as a political and economic entity (e.g. Rowe, 1948) and the fact that Inca rule had come about largely by conquest of neighbouring peoples does not preclude substantial interaction between the various ecological zones in earlier periods (see for example Adams, 1960). This should be emphasised, for in treating a single zone one has to avoid the tendency to think of it in isolation. Equally the word subsistence should not be equated with the notion of self-sufficiency with its implication of limited external contact. It thus becomes less easy to delineate the Peruvian coast in archaeological terms, for we must consider not simply the Pacific littoral but also the desert hinterland and, above all, the canyon-like river valleys which provide natural links with the whole Andean world.

Before embarking on a review of coastal adaptation through time it is appropriate to note the major physical and biological realities of the coastal landscape and, within the limitations of available source material, to attempt to identify environmental changes which could have had a bearing on man's activities. Fig. 1 shows the principal coastal valleys and highland mountain chains, while Fig. 2 gives the location of the archaeological sites mentioned in the text.

COASTAL PHYSIOGRAPHY AND BIOLOGICAL RESOURCES

The Peruvian coastal desert presents an almost lunar landscape, which starts from about 4° south of the equator in southern Ecuador and extends southwards more than 1700 miles into northern Chile. Although there are coastal deserts in other parts of the world, the Peruvian desert is unique in extending so near to the equator. As with other coastal deserts, the basis of the aridity is the presence of cold offshore water, created here by the northward drift of the Peruvian or Humboldt current and augmented by local upwelling. Onshore airflow is therefore cooled in its lower layers, creating a temperature inversion. The air below the inversion is generally chilled to its dew point such that advective stratus may persist for long periods (Eidt, 1968). This is the fog or garua which is responsible for persistent gloom in the southern hemisphere winter. So stable are these coastal atmospheric conditions that even in summer, rain is unknown in normal years. Lima's annual precipitation from fog is 30-50 mm. The fog also substantially reduces temperatures, to

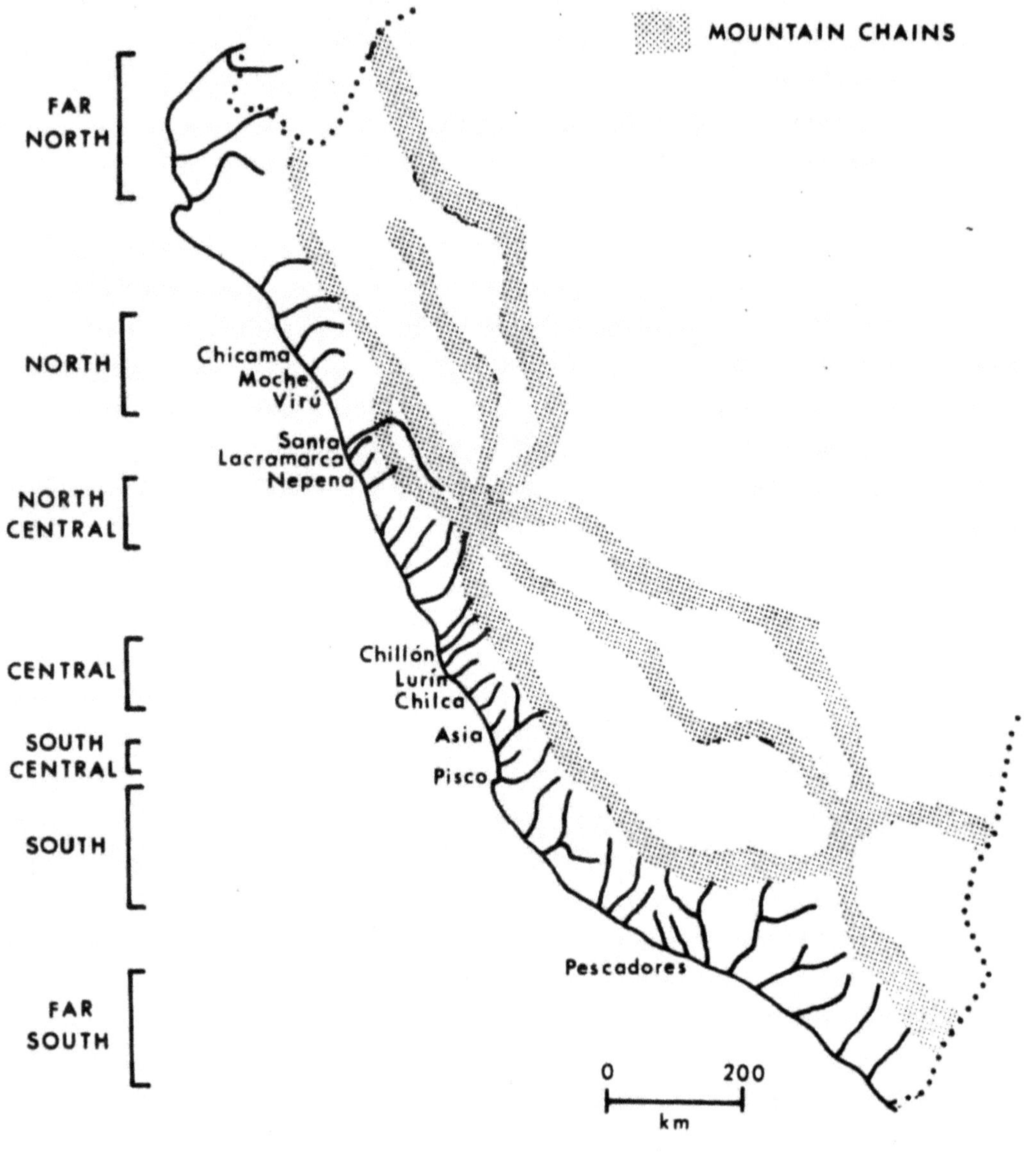

Figure 1. Coastal river valleys and mountain chains in Peru.

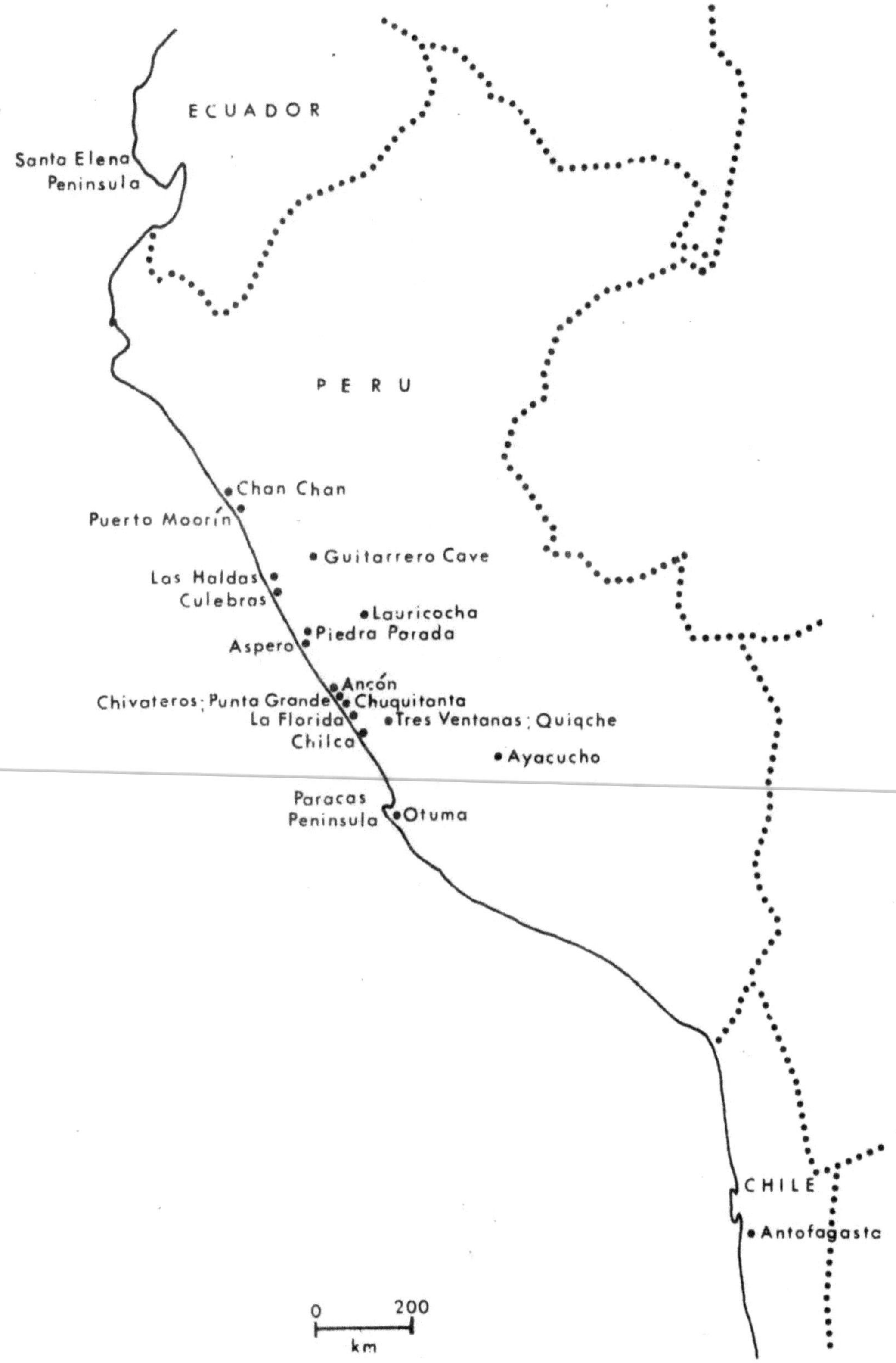

Figure 2. Location of archaeological sites mentioned in the text.

the undoubted benefit of agriculture under conditions where water is frequently limiting. Lima's mean annual temperature of 19°C is therefore distinctly temperate despite maxima of 25-30°C from January to April (Meigs, 1966).

Much of the desert is devoid of vegetation, apart from scattered drought-adapted plants such as Tillandsia and cacti. However, in certain areas where the fog is more dense or lies for longer, condensed moisture from the fog is sufficient to permit the growth of a dense assemblage of plants known locally as the loma or lomas. The fog blanket is generally about 300 m thick and has its base 300-1000 m above sea level, so lomas vegetation is best developed along the foothills and around the canyon-heads of small valleys, a few miles inland from the seashore. The botanical composition of the lomas vegetation varies greatly from one patch to the next, but lomas plants all have certain features in common. Firstly, growth is strictly seasonal, restricted to the foggy months from June to October. Annual plants survive the dry season as seeds in the soil, perennial plants die back to underground tubers, bulbs, corms or fleshy roots. Secondly, the lomas plants do not show the xerophytic adaptations characteristic of the true desert plants. Instead of tough fibrous or spiny foliage which contains little moisture, the lomas plants produce bright green, juicy often highly palatable foliage. In the past, the lomas therefore provided seasonal grazing for animals of all sizes, from snails to deer and camelids. Man would have been attracted to the lomas as a seasonal hunting ground and as a source of edible roots, tubers, bulbs and seeds. However, as well as being seasonal, the lomas are discontinuously distributed, separated by stretches of totally inhospitable desert. Moreover, the extent to which particular patches of lomas develop is notoriously variable from year to year (Parsons, 1970). The lomas therefore constituted a supplementary subsistence resource, not an ecological niche which man could occupy year-round or from one year to the next.

While the ocean water temperature is responsible for the rainless desert on land, it is also the basis of a prolific fishing resource and associated bird life, and this undoubtedly attracted early peoples to the Peruvian coast. The upwelling waters are particularly rich in nutrients and promote an abundant growth of plankton on which in turn feed shellfish, vertebrate fishes and birds. The birds in turn leave excreta (guano), rich in phosphate, which because of the rainless climate has remained as thick accumulations particularly on off-shore islands (Reparaz, 1958).

Parts of the coast consist of rocky bays and headlands, from which line fishing could be practised; elsewhere there are sandy beaches, where fishing nets could be staked out, as is still done today (Lanning, 1967). Both rocky and sandy littorals support a variety of edible shellfish, whose shells are the most abundant constituent of many prehistoric middens. Parmalee and Klippel (1974) have calculated that shellfish alone would not provide sufficient calories to support a population of any size for very long, but there is abundant evidence that prehistoric inhabitants of the coast utilised all the available marine resources, not just shellfish. Fish bones, though less obvious in the middens than mollusc shells, are being recovered in increasing quantities with improved archaeological techniques. Sea birds and sea mammals, including sea lions, elephant seals and sea otters were also caught and eaten. Seaweed has been found archaeologically, not only in middens but also in a mummy bundle and represented on pottery (Towle, 1961), though it is not clear from these contexts what, or how much, use was made of it prehistorically. Today seaweed is processed, dried and sold in coastal cities for use in cooking. It is also fed to animals, particularly pigs, but the only domesticated animals present on the Peruvian coast in pre-European times were guinea pigs (which probably lived mainly on household scraps) and llamas. Seaweed could have been used as a fertiliser or mulch in fields or gardens, though direct evidence of this practice is at present lacking.

The ocean thus provided a subsistence resource which was available throughout the year and from one year to the next. Occasionally, however, this resource fails, due to the activities of El Nino. If the prevailing trade winds weaken or fail, the northward flow of the cold Peruvian current also weakens and the warm Equatorial counter-current sweeps southwards, bringing torrential rains to the unprotected deserts, raising the offshore water temperatures by up to 7°C and reducing their salinity. The upwelling of nutrient-rich subsurface waters also declines, and the combination of diminishing nutrient supply and increasing temperature causes first a decrease in plankton, then drastic effects on the entire marine food chain (Idyll, 1975). El Nino years occur at irregular and unpredictable intervals, perhaps two or three times a century.

Of the fifty or so river valleys which open onto the Pacific coastal plain only about twelve have perennial flow, the majority remaining as dry canyons (quebradas) for varying periods depending on the extent and elevation of their catchments (Reparaz, 1958). The rainy season in the highlands is the summer, which is the sunny season on the coast, while during the winter, when the coast is shrouded in fog, the highlands are dry and sunny. The flow of water in the rivers is therefore greatest, and the valley water tables are at their highest, during the coastal summer when the lomas has dried up. The permanent water-courses provide sufficient moisture to sustain a band of woodland along the river margins. These trees include algarrobo (_Prosopis_ spp.), whose fruits and seeds were a staple food for inhabitants of various dry regions of South America (D'Antoni and Solbrig, 1977). Pacae (_Inga feuillei_) and pepper tree (_Schinus molle_) also have edible fruits, while other trees were used for fuel or construction. Other valley plants have edible seeds, fruits or roots. Cattail (_Typha_), totora (_Scirpus_) and _Cyperus_ have all been found archaeologically (Towle, 1961). In addition, the river valleys supported a varied fauna, affording possibilities for hunting. Valleys without a permanent water supply are, like the lomas, a seasonal resource, but one which complements the lomas in terms of seasonal availability. Valleys with permanent water could be utilised year-round. The availability of water also affects the agricultural potential of the valleys. One crop per year could be obtained from residual moisture on the valley floodplains after the annual floods had subsided. However, once irrigation was available, several crops per year could be produced for as long as the water supply and soil fertility allowed.

The coastal tract itself comprises river flood plains and delta spreads with marine and alluvial terraces at higher levels (Garner, 1959). Between the zone of contemporary beach deposits and these higher surfaces it is common to find salt marsh development, while in the vicinity of major valleys this zone was exploited for water table agriculture (Parsons and Psuty, 1975; R. T. Smith, 1979). Raised shore platforms and fossil cliffs provide evidence of recent coastal uplifts, though such features are not as well developed here as in the zone of stronger westerly winds on the Chilean coast (Fuenzalida et al. 1965). Early shell middens on ancient beaches at Chilca (M. H. Parsons, 1970; Parsons and Psuty, 1975), Otuma (Craig and Psuty, 1971), and on the north coast of Peru (Richards and Broecker, 1963) show that the sea level has fallen relative to the land along much of the Peruvian coast since about 3500 BP. It is therefore evident that major areas of lowland coast utilized for both irrigated and sunken field agriculture are comparatively recent in origin and may not have existed much before 2000 BP.

The soils of the coastal belt vary between those with carbonate and sulphate stratification on more elevated land, to solonchaks on the littoral where water table influence occurs (Beek and Bramao, 1968; Drosdoff et al. 1960; Zavaleta, 1965). Many soils on the coastal plain exhibit multiple layering of different salts including ferruginous banding, sometimes indicative of former rather than present water table influence. In salt marsh conditions sulphidic muds with salt efflorescence are to be encountered while some profiles through beach materials, as at Chilca, reveal thin peaty bands illustrating former salt marsh locations. Most profiles in addition to hallmarks of pedogenesis indicate the transported origin of their parent materials, the littoral zone around river deltas being particularly prone to accumulating successions of sands and silts.

These riverine silts and sands are used for the manufacture of adobe, which was an important building material prehistorically and is still used today. Early houses on the coast seem to have been relatively impermanent structures requiring comparatively little investment of labour. Engel (1976) and Lanning (1967) describe huts from periods before 4000 BP as having walls of sedge or matting supported by canes, willow saplings or whale ribs (presumably derived from animals occasionally stranded on the beaches). Later houses were more permanent structures, with walls of cobblestones obtained from river beds or blocks of coral from the beach, sometimes surmounted by adobe bricks and with wood also used in their construction (Engel, 1976; Lanning, 1967).

CLIMATIC CHANGES

It is as yet difficult to provide an integrated account of climatic changes in western South America during the late Pleistocene and Holocene. Geomorphology, pollen analysis and past distributions of marine animals and land plants have all produced relevant data but comparable studies are not available from all parts of the region and different investigations have sometimes produced results which are difficult to reconcile one with another.

The existence of fossil river valleys in a landscape now swathed in dunes, together with extensive spreads of torrent gravels, are potent indicators of environmental change. Yet vigorous stream flow at the coast most certainly need not imply a formerly moist coastal climate. For example, at the end of the Pleistocene, much of the flow in the rivers would have derived from glacial melt.

Bowman (1916) was first to consider the cutting of the present river valleys and quebradas together with their aggradation as being due to the existence of torrential streams of glacial origin. Fenner (1948) and Garner (1959) both supported this general view. However, glacial melt would have declined progressively almost to the present, with minor reversals, and the main source of water in the coastal rivers is, as it has always been, seasonal precipitation. Many coastal valleys today have only temporary streams and are characterised, in their lower reaches especially, by broad flat floors which have been subject to heavy alluviation, enhanced no doubt by the effects of early removal of woodland at higher levels. But rising flood plains and reduced flows would be expected to lead to shorter overland flow. It is therefore possible that valleys which had earlier been adequately watered became less hospitable in their lower courses so that people gravitated more to the littoral zone and to valleys with permanent streams. This process is endemic to streams in such an environment and the effects of moisture fluctuations in the Andes would be either to alleviate or exacerbate this problem.

Hastenrath (1971) reports on geomorphic evidence for a disproportionately lowered Pleistocene snowline on the west-facing Andes slopes, which in addition to glacially lowered temperatures argues strongly for increased precipitation of westerly origin. Paradoxically there is evidence that at this time upper elevations in the Galapagos Islands were relatively dry compared to the present (Colinvaux, 1972) and that very large areas of tropical rain forest were also relatively dry (Bryan, 1973; CLIMAP, 1976; Gates, 1976; Manabe and Hahn 1977; van der Hammen, 1974; Williams, 1975). The Central and Southern coasts of Peru also seem to have been continuously arid throught the Pleistocene and Holocene (Garner, 1959; Tricart, 1963). During the last cold phase of the Pleistocene, dry conditions must have been widespread across the eastern equatorial Pacific, demonstrated very clearly by the work of Colinvaux (1972) and Quinn (1971) and corresponding to increased development of the Peruvian current and increased upwelling along the equatorial divergence. The moist Andes seem therefore to have been stranded amid aridity and one wonders by what mechanism their humidity was maintained. Beryl Simpson (1975) has discussed this problem and resolved it on the basis of temperature differences between the air over the cold offshore water and the land surface. She argues that, due to glacially altered temperatures in the Pleistocene cold periods, conditions similar to those generating present summer precipitation in the Sierra would have persisted for much of the year. This would certainly produce the conditions for which there is such abundant empirical evidence and argues for the persistence of the sub-tropical anticyclones in glacial times despite computer models which appear to suggest otherwise (Manabe and Hahn, 1977).

Since the end of the Pleistocene, there seem to have been fluctuations in the offshore water temperatures along much of the Pacific coast of South America, probably corresponding to alterations in the strength of the Peruvian current and in the amount of upwelling taking place. This would alter the position of the equatorial convergence. A southward shift in the latter would increase the rainfall on the North coast of Peru, causing the coastal desert to retreat southwards. Further south, warm offshore waters would reduce the amount of fog, bringing about a decrease in lomas vegetation. Conversely, cold waters offshore would create aridity further north, extending at times into the Santa Elena peninsula of Ecuador, and would cause heavy fogs at low levels along much of the Peruvian coast, with correspondingly greater development of the coastal lomas.

There is evidence that the North coast of Peru was comparatively humid around 8000 BP. Mangroves then extended 200 miles south of their present limits with denser vegetation than at present inland (Richardson, 1973; Lanning, 1967). This particular phase of warmer offshore waters seems to be detectable along a considerable length of the Pacific coast, for remains of fish which today do not extend south of the coastal regions of Peru and Ecuador have been recovered from levels dating around 9500 BP from Quebrada Las Conchas, near Antofagasta in Chile (Llagostera M., 1979).

Later levels of sites in the Antofagasta area, dated around 5350 BP, yielded abundant remains of the mollusc <u>Choromytilus</u>, which today is found only in the far south of Chile and which therefore indicates offshore waters colder than at present (Llagostera M., 1979). Pollen data from Chile cited by Llagostera M. (1979) also indicate a cool humid period from 6500-4500 BP. In northern Peru, shells of mangrove-dwelling molluscs, present in abundance in Siches phase sites (8000-6000 BP), are absent from the succeeding Honda phase sites (5101 $\pm$ 100 BP) although Honda sites, like Siches sites, are located on the coast and their inhabitants collected molluscs from the intertidal zone (Richardson, 1973). This absence of mangrove molluscs suggests that the mangroves had retreated north, possibly in response to a cooling of the coastal waters and a corresponding reduction in coastal precipitation.

However, slightly warmer offshore waters apparently soon returned, indicated by bones of the warm-temperate fish <u>Trachurus</u> in occupation levels dated 4000-3550 BP near Antofagasta (Llagostera M., 1979). Further north, data from the Santa Elena peninsula of Ecuador suggest a pluvial period from 5000-3600 BP, succeeded by an arid period between 3600 and 3000 BP. These data also correlate with successive pluvial and dry periods shown in pollen profiles from various localities in Colombia (Sarma, 1974).

Paulsen (1976) has visualised climatic fluctuations in coastal Ecuador as indicating parallel wet and dry phases in the Sierra. While it is tempting to fit all areas into a unified pattern of general climatic change, there is considerable need for caution because of the complex nature of the factors controlling the local climates. In the north of Peru and in Ecuador, rainfall patterns will relate more clearly to the equatorial convergence but in the central and southern highlands the influence of the equatorial convergence is very much reduced so that one is left simply with variations in temperature of offshore water as the most obvious factor affecting rainfall at high levels. More rain in the highlands would have meant more and longer flows of water in the coastal river valleys, which would presumably make these suitable habitats for more months of the year both for man and for the animals he hunted. Higher water tables might likewise encourage the developing techniques of plant cultivation.

Climatic changes since the end of the Pleistocene are therefore likely to have affected man's exploitation of the biological resources of the coast in three main ways. Firstly, shorter overland flow of some at least of the westward-draining rivers would result from heavy alluviation brought about initially as glacial melt water drained away and later as human clearance of natural tree cover increased erosion. Some of the coastal quebradas would therefore become dry in their lower reaches, necessitating a shift of population to better-watered valleys or to the coast itself. Secondly, fluctuations in offshore water temperatures would bring about expansion or retreat of the lomas vegetation, affecting the size of human or animal populations that this resource could support. Thirdly, fluctuations in the amounts of rainfall in the highlands, possibly correlated with changes in offshore water temperatures, would affect the flow of water in the rivers and the height of the water tables in the coastal valleys, with consequences for both natural vegetation and planted crops.

SUBSISTENCE CHANGES THROUGH TIME

Given the generally inhospitable environment of the Peruvian coastal plain, it would not be surprising to find that human occupation of the coast is more recent than that of the highlands. Early dates for man in Peru are as controversial as those from other parts of the Americas. Two which have been widely quoted are around 19,000 BP for the Paccaicasa complex from the Ayacucho basin of highland Peru (MacNeish, 1977) and around 14,000 BP for the bottom of the Red Zone at the quarry site of Chivateros on the coast (Lanning and Patterson, 1967). However, Lynch (1974) has produced some cogent reasons for regarding both of these occupations, and also the Ayacucho complex which succeeds Paccaicasa in the highlands, with considerable reservations. The earliest sites which are stratigraphically sound, have an acceptable stone industry and have been dated by C14 analysis of associated charcoal are Guitarrero Cave (12,610 ± 360 BP), Tres Ventanas and Quiqche Caves (10,080 ± 170 BP and 9990 ± 200 BP), Lauricocha (9575 ± 250 BP) and Jaywamachay Cave (9510 ± 160 BP; 8695 ± 140 BP) (Lynch, 1974). All these are highland sites. On the coast, sites dating to around 9000 BP occur virtually all along the arid plain: Quebrada Las Conchas, near Antofagasta in what is today Chile (9400 ± 160 BP, 9680 ± 160 BP; Llagostera M., 1979); village 96 on the Paracas peninsula (c. 9000 BP; Engel, 1976); the Arenal complex from lomas sites around Ancón (9000-8000 BP; Lanning, 1965); the Siches phase of northern Peru (starting date c. 8000 BP; Richardson, 1973) and the Vegas period on the Santa Elena peninsula of Ecuador (starting date 8850 BP; Sarma, 1974). Regardless of what starting point one accepts for Peruvian prehistory, it does therefore seem that extensive human settlement on the coast occurred a millennium or so later than occupation of the highlands.

On the coast, it is only after about 9000 BP that there are sufficient data to enable tentative reconstruction of subsistence patterns. Preservation conditions are generally good, and both lomas and shore are less intensively occupied now than they were, at least during certain periods of the past, so archaeological surveys of these zones probably do give a reasonably accurate picture of the number and distribution of sites of different periods. The river valleys, however, have been, and are, intensively cultivated, resulting in loss or burial of many sites in this zone.

Archaeological work has concentrated on the Chillón, Lurín and Chilca valleys of the Central coast, but the more limited data from south and north of this area fit the same general pattern. Three different modes of exploitation of the available resources can be distinguished, and on the whole succeed one another chronologically. During the Lithic Stage (c. 9000-4500 BP) populations were seasonally nomadic, moving between lomas and river valleys and using the sea as a subsidiary resource only. During the Cotton Preceramic (4500-c. 3800 BP) marine products formed the bulk of the diet, and cultivation of plants became widespread. Following the introduction of pottery (c. 3800 BP onwards) much of the coastal population moved inland and became full-time farmers.

1. Lithic Stage (Preceramic Periods I-V) c. 9000-4500 BP

Cultures from this period are best known from sites in the lomas. Since the lomas is a strictly seasonal resource, these sites are presumed to represent winter camps of nomadic or semi-nomadic peoples. Artifacts found indicate a hunting and gathering economy. They include projectile points, grinding tools and cobble flakes, which might have been used as sickles. Both grinding tools and cobble flakes are rare or absent at first but increase in frequency from about 7000 BP onwards (Cohen, 1977). Detailed reports of faunal and floral remains from lomas sites are not available, probably in part because of the difficulty of identifying roots, tubers or seeds from the poorly known lomas flora. According to Engel (1976), coprolites from lomas sites contained remains of grass seeds, cactus fruits and begonia tubers. Lanning (cited in Cohen, 1977) found that the amount of seed in refuse of lomas sites increased through time until, at the end of the Lithic stage, the quantities of grass seed were such as to suggest systematic harvesting and storage. This increasing reliance on seed correlates with the increasing frequency of grinding tools and cobble flakes, and with a decrease in the bones of land vertebrates. The reduction in protein derived from deer and other animals was in part compensated by greater use of marine resources, shown by increasing amounts of shell in the refuse.

Summer camps of the lomas dwellers were presumably located in the river valleys, which would have their maximum flow of water during this season. Such a summer camp has been reported from the Chilca valley (Patterson, 1971a). Here subsistence involved primarily plant collecting and hunting in the valley-floor habitat, with shellfish providing a secondary resource. The site is presumed to have been occupied during the summer only because, although it is located near an extensive lomas, no lomas products were found among the refuse.

In the Lurín valley, a Lithic site has been excavated which was apparently occupied year-round, since marine, riverine and lomas products all appear in the debris (Patterson, 1971a). Unlike the seasonal camps, in which hunting and gathering predominated, at this permanent site fishing and shellfish collecting were most important, with lomas and river valley providing seasonal supplements to a basically marine diet. On the Paracas peninsula on the South coast, village 96 seems to have followed a similar marine-orientated mode of subsistence. Excavation of this village has produced the earliest fishing net yet known, dated to about 9000 BP and made of a fibre other than cotton (Engel, 1976). Engel (1976) considers that fishing must have been done from the shore, possibly by fishermen swimming with nets in their hands as shown on the painted pottery of later South coast cultures; possibly by staking out nets on a sandy shore (Lanning, 1967). There is no evidence of rafts or any other sort of boat in Andean South America as early as this. Further south still, in Chile, analysis of fish bones from Quebrada Las Conchas has shown that the majority of bones represent bottom dwellers and one, Cynoscion, must have been caught by net since it does not feed on anything which could be used to bait a hook (Llagostera, M., 1979). Fish hooks made of shell, bone or bent cactus spines were, however, used during the Lithic stage in both Peru and Chile.

Two plants often cited as cultivated appear on the coast during the Lithic stage: the bottle gourd (Lagenaria siceraria) and one or more species of squash (Cucurbita). Bottle gourd fruits were very valuable to pre-pottery cultures as containers and also served as fish-net floats. The squash has been variously identified as C. ficifolia (Cohen, 1977) and C. ecuadorensis (Cutler and Whitaker, 1969). It is sometimes claimed as the first food plant cultivated on the coast of Peru. This may be so, but the plant is likely to have been grown for its edible oil- and protein- rich seeds rather than for its fruit. Wild cucurbits such as C. ecuadorensis, and also C. ficifolia, have comparatively little flesh, which is coarse and stringy and, in the wild species, often bitter. Seeds, on the other hand, seem to develop bitter principles only on germination (Whitaker and Davis, 1962), so seeds even of wild cucurbits would be edible. Both C. ecuadorensis and C. ficifolia have hard-shelled fruits and it is conceivable that they were used, like Lagenaria gourds, as fish-net floats and containers rather than as food. Cucurbits in general are capable of extremely rapid growth (providing they have access to adequate water) and can smother competing weeds. They can grow and maintain themselves with minimal, if any, attention from man, as witnessed by the observations of one of us (B.P.) on feral C. ficifolia growing on scrub-covered slopes in the Andean highlands and feral Lagenaria on lake margins and along ditches in the dry lowlands of north-east Brazil. Cucurbita and Lagenaria may both have been introduced by man to the Peruvian coast, but this does not presuppose much agricultural skill and possibly not even deliberate planting.

Lithic sites on the North coast also probably represent seasonal camps, located in the river valleys during the summer and on the coast during the winter (Lanning, 1967). No projectile points have been found and hunting does not seem to have been important. Coastal camps include abundant remains of mangrove molluscs, fish and bones of sea mammals, while mortars may indicate some use of wild seeds. Summer season sites have not been excavated, so it is not clear what were the main food resources during this season.

On the South coast, too, there is evidence of winter campsites in the lomas, but the summer camps seem to have been in the upper headwaters of the valleys. Hunting was of major importance, to judge from the numerous projectile points, but the artifacts imply that there was little use of plants. Lanning (1967) suggests that the extensive and luxuriant lomas of the south supported a denser population of game than occurred elsewhere on the coast, and the Lithic hunters followed this game on its seasonal rounds from lomas to highlands and back again.

At the Lithic level of technology, therefore, adequate subsistence could have been achieved by sedentary life in a strategically situated site which had access to both marine resources and to the more seasonal resources of lomas and river valleys. The alternative was seasonal nomadism, with temporary camps shifting from lomas to river valley and back again. It is possible that nomadism over longer distances also occurred. The complementary seasonality of coast and highlands produced a migration of game from winter grazing grounds on the lomas up to the highlands for the summer rainy season and down to the coast again when pastures in the highland basins dried up. Man may have followed the game animals on this seasonal round. There were certainly contacts between coast and highlands during the Lithic Stage, as evidenced by the presence of highland obsidian in coastal sites (MacNeish, 1977), presence of blue mussel shells in caves at 3600 m 60 miles from the sea in the Chilca valley (Engel, 1970, 1976) and typological similarities in projectile points from sites in the highlands around Ayacucho and lomas campsites around Ancón (Cohen, 1977; MacNeish, 1977). Whether this represents physical movement of people or a well-developed trading network is, however, not yet clear.

2. Cotton Preceramic (Preceramic Period VI) 4500-c. 3800 BP

During the Cotton Preceramic man abandoned much of the lomas and marine products became of primary importance in subsistence. Marine resources are available throughout the year, so the nomadic or semi-nomadic way of life could be discarded and permanent settlements built on the coast, often at the valley mouths. Some of these settlements are very large, for example Chuquitanta (El Paraíso) in the Chillón valley, which extends over 50 hectares and presumably housed a sizeable population. The Lithic pattern of seasonal nomadism does, however, seem to have persisted along the South coast (Lanning, 1967).

There were probably several reasons for the abandonment of the lomas and the corresponding changes in settlement patterns. Lanning (1963) suggests that climatic change led to a gradual desiccation and retreat of the lomas. He observed that naturally-occurring beds of land snails are located in areas where there is no vegetation today and hence he considered the snail beds to indicate past areas of lomas. However, Moseley (1975) points out that modern land snails extend well beyond and below extant areas of lomas, and Parsons (1970) argues convincingly that temporary expansion of the lomas during El Nino periods would leave relict beds of snails in what, in non-Nino years, is arid desert.

Other evidence for climatic changes which would affect the lomas has already been discussed. Warmer offshore waters around 4500-3500 BP could have reduced the extent and duration of the coastal fogs with deleterious effects of the lomas vegetation. This might have triggered a switch in subsistence away from the lomas and towards greater use of marine resources. However, climatic change is unlikely to be the whole explanation. In particular, hypotheses of a reduction in the lomas do not fit very well with Cohen's (1977) observation that the distance of the lomas sites from the river valleys increases gradually throughout the Lithic, suggesting that the nearer lomas had been over-exploited and the population was having to travel further for its winter resources. Cohen (1977) argues convincingly that this changing distribution pattern of lomas sites would not be produced by climatic change. He and Patterson (1971b) see the switch away from the lomas to other forms of subsistence as a necessary response to an increase in population above the carrying capacity of the lomas.

Patterson (1971b) estimates this population increase as sixfold along the Central coast and suggests that it might have been triggered by dietary improvements late in the Lithic Stage, particularly increased use of marine protein and use of seeds instead of high-starch, low-protein vegetative storage organs of plants. Tuberous or bulbous plants may well have been exterminated before plants with edible seeds if man was indeed over-exploiting the lomas. Plants with underground storage organs are perennials, hence often take longer to reach reproductive maturity than seed-reproducing annuals. Their seed output per year is also usually lower than that of annuals, and these two factors mean that the ability of their populations to compensate for loss of individuals through over-harvesting by man is more limited. Furthermore, since seeds are smaller than tubers or bulbs and more readily dispersed, more seeds will escape the harvesting process and survive to regenerate the population. Over-exploitation of the plant resources of the lomas might therefore have forced man to rely increasingly on seeds as harvests of edible storage organs diminished. Eventually man abandoned the lomas altogether.

Although marine products were the primary subsistence resource during the Cotton Preceramic, cultivated plants are more abundant and include a greater range of species than formerly. Lithic Stage sites have yielded only gourds and squash, and possibly _Phaseolus_ beans (Pickersgill, 1969). The most frequent cultigens in sites of the time period 4500-3800 BP are cotton and gourds, both of which are associated with fishing technology and therefore fit with the marine orientation of these cultures.

The cotton present at the beginning of the Cotton Preceramic differs little if at all from modern wild _Gossypium barbadense_, which still grows in arid coastal parts of the Santa Elena peninsula and the extreme north of Peru (Stephens and Moseley, 1974). The archaeological specimens show a gradual increase in seed size (which would result in increased yield of fibre) and in fibre diameter through time. This suggests _in situ_ domestication of _G. barbadense_ in coastal Peru or Ecuador. Demand for cotton increased rapidly, and it was soon used not just for making fishing lines and nets, but as the preferred fibre for increasingly elaborate textiles which Moseley (1975) considers symbolised differences in wealth and status within the population as well as serving as the main medium of artistic expression. Wild cottons are all perennial plants of very dry habitats and the early cultivated cottons cuuld probably have survived quite satisfactorily on the floodplains of the coastal valleys.

Other cultivated plants utilised during the Cotton Preceramic include common beans, lima beans, _Canavalia_ beans, chili peppers and the root crop achira (_Canna_) (Pickersgill, 1969). More recently other root crops have been identified from preceramic sites. These include sweet potato, manioc, the highland root crops oca, ullucu and Irish potato (Martins, 1976) and yam bean (Grobman and Bonavía, 1978). These crops are less drought-resistant than cotton, but most are annuals or can be grown as annuals. They were presumably planted in the river valleys as soon as possible after the rivers subsided from their annual crests and matured on the residual moisture left in the soils after flooding.

In the absence of quantitative data or full site reports it is difficult to evaluate the dietary importance of food plants compared to marine products. Many large settlements of this period, such as Chuquitanta, Aspero, Piedra Parada and Culebras, were located near the mouths of rivers with access to reasonable areas of land suitable for floodwater farming (Moseley, 1975). Unfortunately no middens associated with these sites have yet been fully excavated. However, it seems likely that cultivated plants would have been more important in the diet at these sites than at the more intensively studied sites around Ancón (Moseley, 1975) and at Las Haldas (Matsuzawa, 1978) where there was no nearby land suitable for cultivation, but where large communities nevertheless thrived.

Very little work has been done on inland river valley sites away from the coastal plain, but it is in this zone that man probably first became wholly dependent on cultivated plants and consequently developed the technological improvements presaging the switch from fishing to farming on the coast itself after 3800 BP. According to Patterson (1971a), these up-valley sites are located where river water, fuel, building materials and arable land would be available year-round. Away from the coastal plain the rivers run in deep channels and floodplains are virtually non-existent, so to bring adequate land under cultivation it would be necessary to develop a system of canal irrigation. The dramatic relief of the coastal foothills, responsible for the deep down-cutting of the river channels, means that only short lead-off canals would be needed to get water from the rivers, and natural slopes would facilitate distribution of irrigation water over the cultivable land (Patterson, 1971a; Farrington, 1974; Moseley, 1975). In contrast, on the coast there is little natural head of water and lengthy canal systems were necessary to conduct water from the rivers to the fields. For example, in valleys such as the Chicama, lateral canals 24-36 miles long were constructed along one or both banks of the river (Engel, 1976). Further distribution and drainage canals were also needed to compensate for the lack of slope in the cultivated fields.

During the Cotton Preceramic, then, the previously nomadic way of life with its seasonal round from lomas to river valley was abandoned and replaced by permanent settlements relying on the non-seasonal products of the ocean (fish, shellfish, seabirds and sea mammals) and/or plant cultivation. The floodwater farming probably practised on the coast at this period would yield only one crop per year, but many of the plants grown produce a storable surplus, for example, dried beans, maize kernels, sweet potatoes and manioc tubers (the latter stored _in situ_ in the ground and harvested as necessary). These surpluses would enable the population to survive until the next harvest. A corollary of abandonment of the nomadic life-style was the loss of community self-sufficiency and the beginning of specialisation in food production. Thus, coastal sites such as Las Haldas which lacked access to cultivable land received agricultural produce through trade while marine shellfish were traded to the predominantly agricultural up-valley sites. These trade networks must also have involved agricultural communities in the highlands, since the potatoes, oca and ullucu reported from Ancón (Martins, 1976) are highland crops which would not have grown successfully on the coast (Hawkes, 1967; León, 1964). Although a wide spectrum of crop plants was available and would have provided an agriculturally efficient cropping and rotation system as well as a nutritionally balanced diet, agriculture spread only slowly on the coast and remained a subsidiary, presumably not particularly desirable, food procurement system throughout the Cotton Preceramic.

3. Ceramic periods c. 3800 BP onwards

About 3800 BP many of the settlements of the Cotton Preceramic were abandoned. Sites affected include the small farming villages of the South coast around Chilca as well as the large, presumably ceremonial, centres on the Central coast. Sites with different subsistence orientations, such as Punta Grande, with its marine specialisation, and Chuquitanta, with its access to suitable land for floodplain farming, were likewise abandoned. Engel (1976) considers that the South coast was probably deserted for a few centuries after this. On the Central coast, large new ceremonial centres were constructed (Lanning, 1967). A comparatively small fraction of the population remained on the coast and continued the previous pattern of marine-based subsistence, but most of the population moved inland, probably into smaller and more dispersed settlements, and became full-time farmers (Moseley, 1975).

As with the earlier subsistence shift from hunting-and-gathering to fishing, it is not easy to identify precise reasons for the change from dependence on fishing to dependence of agriculture. Climatic change per se seems unlikely, since there is no evidence of any marked alterations in temperature or rainfall at this time. However, two widely separated localities, the North coast (Richards and Broecker, 1963) and Otuma on the South coast (Craig and Psuty, 1971) provide evidence of a period of tectonic uplift around 3600-3000 BP. Beaches were raised up to 9 m on the North coast. These disturbances must have affected both fishing and farming economies, particularly since, around Otuma at least, the uplift occurred so suddenly that beds of molluscs were stranded in situ, retaining their original patterns of zonation on the shore (Craig and Psuty, 1971). Raising the land results in an effective lowering of the water table, at least with regard to the land surfaces which existed before uplift. This does not seem particularly conducive to agriculture.

Patterson (1971b) suggests that the population of central Peru increased fivefold between 4300 and 3750 BP. However, population pressure does not seem to have resulted in over-exploitation of the sea's resources. The stranded lagoon at Otuma, and the Bay of Ventanilla (drained by a later uplift), still contained vast beds of edible molluscs with little evidence of over-fishing (Craig and Psuty, 1971; Moseley, 1975).

Abandonment of previous settlements, introduction of new technology (pottery) and a change in the primary method of food procurement might all suggest conquest of central Peru by an alien group which imposed its own way of life. However, according to Lanning (1967), regional differences intensified during this period and the early Peruvian pottery itself represents at least three fundamentally different traditions. This would seem to eliminate conquest as a cause of a change in subsistence.

It therefore seems likely that the switch to agriculture resulted from technological innovation rather than environmental change, population pressure or conquest. The innovation most likely to have brought about increase in the amount of land and quantity of plants cultivated, and to have caused dispersal of inland settlements over wider areas, is the introduction of canal irrigation. Irrigation systems are not easy to date accurately, and there is unfortunately no direct evidence of irrigation on the coast as early as 3800-3000 BP, though the locations of settlements such as La Florida strongly suggest that irrigation was practised (MacNeish, Patterson and Browman, 1975; Moseley, 1974).

Pottery itself is an innovation important to agriculture insofar as it must have increased the acceptability of some of the products of cultivation by greatly facilitating the preparation of foods which require long periods of boiling. This applies particularly to the various beans, for which prolonged cooking is necessary to break down trypsin inhibitors, phyto-haemagglutinins, flatus factors and cyanogenetic glycosides, some or all of which are present in the raw seeds. Preceramic cooking techniques involved roasting, or boiling by adding heated stones to water in a container such as a gourd which could not be heated directly. Large quantities of fire-cracked stones in Ceramic period sites indicate that these techniques persisted after the introduction of pottery even though most of the early pots are in fact cooking vessels (Lanning, 1967).

Other innovations were added during succeeding periods without altering the established pattern whereby communities were engaged primarily in either farming or fishing, and were linked through trading networks with other coastal and even highland communities into an internally interdependent but overall self-sufficient unit. Totora reed boats were introduced on the north coast shortly before 2000 BP (Moseley and Mackey, 1972) and enabled exploitation of the previously inaccessible deep-water marine fauna. In Chile, Llagostera M. (1979) has used presence of conger eel in occupation debris as an indicator of use of rafts, and has obtained a very similar date for the start of deep-water fishing. The foot plough or taclla seems to have been an even later, possibly Inca, introduction to the coast. It made possible the cultivation of low, wet flood-plains whose clearance required some form of heavy clod-cutter (Farrington, 1974). A uniquely coastal innovation was development of the techniques of sunken field cultivation, which merit more extensive discussion.

WATER TABLE AGRICULTURE

Early field systems in Andean America are widespread yet it is comparatively recently that they have received attention in the literature. One of these forms is the sunken field, recently investigated by Jeffrey Parsons (Parsons, 1968). Parsons' paper triggered immediate responses by two authorities on the desert coast, Moseley (1969) and Rowe (1969), while further research has led to a greater understanding of what these fields were used for, how they may have been managed, their actual antiquity and their relationships, both in time and space, to other forms of subsistence (Parsons and Psuty, 1975; Kautz and Keatinge, 1977; Smith, 1979; West, 1979).

Sunken fields generally occur in the topographically low zone slightly inland of the active littoral belt although the environmental prerequisite, a water table which is very near to the ground surface, is achieved sometimes more than a kilometre inland, as in the Pisco area. The main sites are by the mouths of some ten or so valleys, only a few of which have perennial streams. These valleys are, from north to south, the Chicama, Moche, Santa, Lacramarca, Nepena, Chilca, Asia, Pisco and Pescadores. The total area occupied by sunken field sites amounts to about 3500 ha and the total cultivated surface to about 2000 ha (Parsons and Psuty, 1975). Given the propensity of these excavated sand plots to become obscured and destroyed with time, these must surely be minimum figures. Indeed Rowe's discussion of water table agriculture (Rowe, 1969) indicates that depressions which were naturally filled with fresh water were also exploited and this gives cause for wondering whether the sunken field was developed first in areas where such natural 'oases' occurred. Known at Chilca on the Central coast as mahamaes and on the North coast as pukios (puquios) these agricultural plots bear some resemblance to the walk-in wells of the North coast (Moseley, 1969, p. 486), Ecuador and highland Mexico (Flannery et al., 1967). Rowe (1969) finds that in most of the early literature the word hoya (Spanish for pit or depression) was used for these features, this word being used today in the south of Peru.

The fields themselves were originally excavated in the sands and silts of the coastal zone until a saturated zone was reached. The field surface can thus be up to 2 metres below the original ground surface. Characteristically each field is surrounded by steep-sided midden debris consisting of excavated sand and silt and periodic crop residues which survive here in desiccated form. Often the field basins are walled by blocks of salt-crusted soil. The sunken basins may be single or occur collectively. They may be small and squarish or long linear features extending for up to 100 m or so. Basins may also form a continuous reticulate pattern. Usually there is some alignment involved in the layout of the fields suggestive of systematic planning though the original topography may have suggested or even dictated these trends in some instances. The amount of field excavation relates to the depth at which a humid zone could be exposed but since virtually all these fields are now abandoned we can only surmise the depth to which aboriginal peoples considered it prudent to excavate given the risks of salinization. Ground water tables have fallen in the course of time due to slow coastal uplift and reduced river discharges, while water abstraction, from wells in particular, has meant the deepening of plots particularly in zones of most intensive and prolonged use. The view has been expressed elsewhere (Smith, 1979) that this drying process led to production of smaller plots in order to reduce the work involved in lowering field surfaces. Simple layouts and isolated basins are therefore interpreted as signifying a limited period of use.

The sunken gardens produced a wide range of food and fibre. Middens associated with sunken fields at Chilca reveal maize, beans, cotton, gourds, squash, peanuts, coca, chili peppers, manioc, totora and the tree fruits, lucuma, ciruela del fraile (Bunchosia) and cereza (Malpighia) (Pickersgill, unpubl.), while pollen from field soils confirms most of these together with cattail, sedge (totora perhaps) and possibly quinoa, though the latter has so far not been recovered from midden deposits (R. T. Smith, 1979). West (1979) has found pollen of Solanaceae at Puerto Moorin in the Virú valley together with records of maize and beans. This total range of crops argues for the substantial adaptability of the system albeit involving suitable choice of sites and quite possibly differential excavation. It indicates that, for short periods at least, these fields were viable notwithstanding the salinity hazard. Fish heads (anchovetas) were placed as fertilizer in the same hole as seeds, this practice being general along the Peruvian coast and apparently more widespread than the use of guano (Rowe, 1969). Guano was, however, used by Chan Chan farmers on the North coast, who are reputed to have had it brought as much as 435 miles from the Chincha Islands near Pisco (Reparaz, 1958). Such use of marine resources, particularly fish, as fertilizer underlines the primary dependence of coastal people on the sea and the role of the sea in the development of agriculture.

Sunken fields would have afforded all year round production compared to areas dependent on seasonal flood waters while the locations chosen for sunken fields, generally near to valleys, guarantee a supply of moist mud and sand to make adobe for building, another vital consideration in the context of coastal adaptation. Finally, and clearly of immense importance, the coastal fogs must have played a crucial role in rendering viable the exploitation of the water table in the first place. This arises from the reduction of evapotranspiration which would greatly retard the rate of salinization and thus lengthen the period over which field surfaces could remain in use.

The existence of coastal sunken fields requires explanation, since they imply a considerable labour input. Those at Puerto Moorin may be as early as 2000 years BP (West, 1979) while the ones at Chilca are generally no earlier than 1300 BP (Parsons and Psuty, 1975) and those at Chan Chan seem mainly to have been constructed and used in Imperial Chimu times around 900-1000 BP (Farrington, 1974). It is thus clear that sunken field agriculture occurred far too late in the pre-hispanic sequence to have exercised any formative influence on the development of coastal agriculture.

It has been suggested that sunken field agriculture arose through population pressure and excessive demands on irrigation systems or on account of difficulties experienced with the latter (J. R. Parsons, 1968). Clearly no single argument need account for the development of mahamaes in every locality. It is only at Chan Chan that these fields are associated with a major urban centre (maximum population estimated at 200,000 persons) large enough for us to envisage population 'pressing' on available resources. Moseley (1969) argues that it is improbable that difficulties with irrigation had any influence on the spread of mahamaes but he neglects to consider the consequences of environmental change which, as stated earlier, would tend in the less watered valleys to favour such coastal exploitation in the face of siltation and reduced flows up valley. Furthermore it is likely that in the centuries around 2000 BP, more and more low coastal plain was revealing itself as a new environment for exploitation. Engel (1976), who has made detailed investigations in the Chilca valley, finds that earlier peoples used well water to irrigate the valley floor but these groups appear to have left about 1800 BP. This may indeed be evidence of the valley siltation problem already mentioned.

It is clear that sunken fields are merely supplemental to irrigated agriculture in some localities while in areas lacking irrigation opportunities sunken fields represent the sole method of farming, unless of course water could be raised from wells. The main argument against the latter is that if an efficient system for raising large quantities of well water had been devised there would be little point in making such deep or extensive excavations. Whether mahamaes were first developed in areas already with irrigation farming or were invented by resourceful people making the most of an area totally lacking surface water is impossible to say at present except that the argument based on observation of water-filled depressions, mentioned earlier, is a persuasive one. What seems more to the point is why they were made and, in particular, why such a large area was developed at Chilca when there seems to be so little evidence of settlement coeval with the fields. Although early settlement may have been obliterated by the development of the Spanish town of Chilca the main adobe structures peripheral to the sunken field zone are, according to Engel and other workers (Parsons and Psuty, 1975), earlier than the fields, with radiocarbon datings around 2000 BP compared to 1300 BP for the fields. Engel (1976) considers the sunken field people to have brought new influences to the area as evidenced by Tiahuanaco-style pottery (locally referred to as Puerto Viejo type) and this permits an explanation other than simple development by coastal peoples for their own purposes. Around 1300-1200 BP the Huari people were expanding their influence in the highlands immediately east of the Central coast and one may therefore be justified in associating Tiahuanacoid pottery styles with this expansionist phase. Parsons (in Parsons and Psuty, 1975) also considers that Huari influence explains the Chilca mahamaes as signifying local population expansion. There are two principal possibilities: Huari people came to the coast either in search of new lands to settle or simply to exploit the coast in order to meet the demand for food and other materials in the highlands. The first would imply a larger resident population than the second though either would explain the fact that virtually the entire system of fields at Chilca appears to have been planned as a unit. Use of fish as fertiliser must again argue for an effective fishing industry so that the Huari Empire stood to gain both fish and coastal crops from its expansion westward.

Sunken fields were thus used at various times on the Peruvian coast from about 2000 BP onwards. They were sometimes related to local settlement specialisation as in the Virú valley (West, 1979), sometimes provided a supplement to canal-based irrigation farming as at Chan Chan, while at Chilca the system was probably developed by outside influence. The latter may have functioned to cater for highland population overspill but more probably Chilca was set up as a self-sufficient community whose primary role was the exploitation of marine resources.

SUMMARY AND CONCLUSIONS

The desert coast of Peru provided an environment for prehistoric man which is without an exact parallel anywhere else. Its marine resources are among the richest in the world and even today the waters off Peru and northern Chile produce more than one-fifth of the world's commerical fish (Moseley, 1975). In contrast, the desert and the river valleys which cross it are only seasonally productive. The archaeological record shows that man devised various strategies for subsisting satisfactorily in this environment. The first involved exploitation of the seasonal resources of lomas and river valleys and necessitated migration between summer valley and winter lomas camps. The sea, which was equally accessible from both the lomas and the river valleys, was used year-round to supplement the diet. Cultivated plants may also have provided a minor food supply. The second subsistence system relied primarily on marine products. The nomadic way of life was abandoned, permanent settlements were built along the coast, and agriculture became more important, though still only a subsidiary source of food. The third food procurement system involved full-time farming. Initially man utilised the seasonal floods in the river valleys to raise one crop a year and relied on a storable surplus to support the population between harvests. Later, irrigation made it possible to grow crops year-round, at least in valleys with a permanent supply of water.

There seems to have been little interaction between groups following the first and second subsistence strategies, since lomas campsites were abandoned at about the time that permanent settlements of full time fishermen became widespread. However, exponents of the second and third strategies have co-existed probably since the Cotton Preceramic when specialised communities of farmers or fishermen first appeared (Moseley, 1975). These specialised settlements were linked by exchange networks into self-sufficient communities, a pattern which persisted up to the Spanish Conquest (Rostworowski de Diez Canseco, 1977) and still continues today.

The earliest occupation of the coast was by semi-nomadic hunters and gatheres. Permanent, marine-based settlements came later, while dependence on agriculture developed later still. This sequence, from nomadism to sedentism and, ultimately, agriculture, fits Sauer's (1952) theories on the cultural context in which man would have started to cultivate plants. However, development of agriculture was seldom as straightforward as Sauer implies. Certainly study of the archaeological record from the Peruvian coast suggests that it cannot be read as a simple, still less as an inevitable, progression from hunting and gathering to farming.

Hunting and gathering may have been the only way of life open to the earliest occupants of the coast. Marine shellfish can be gathered by people at even the most primitive level of technology, but Parmalee and Klippel (1974) have calculated that man could not subsist on molluscs alone for any length of time. Seabirds and sea mammals could be shot, speared or harpooned using weapons similar to those used for hunting land game, but vertebrate fish were probably the most important marine source of calories and protein. Freshwater fish can be caught in various ways, including poisoning or trapping, but these latter methods are not applicable to fish living in the open ocean, most of which must be caught by line or by net. Fish hooks and nets have been found in Lithic sites, but cotton is not known until later in the Preceramic. Whatever fibre was used to make Lithic lines and nets, it is unlikely to have been as strong or capable of producing such a fine-meshed net as cotton. Certainly once techniques for making thread and cloth from cotton fibre had been invented, cotton spread rapidly all along the coast—as a cultivated plant, since lint-producing cottons occur wild only in the extreme north of Peru and demand for fibre very soon outstripped the quantities that could be supplied by wild plants. Moseley (1975) sees the Lithic stage as providing the necessary time for a slow build-up of marine technology which enabled coastal populations to switch from land-based hunting and gathering to a marine-based subsistence around 4500 BP. Cotton was probably an important element in the development of this marine technology, and it is something of a paradox that development of a fishing economy should be linked with prior establishment of agriculture or at least of the techniques for cultivating cotton and gourds. Peruvian coastal agricultural is in turn unusual, if not unique, in its early emphasis on two non-food crops which provided important raw materials for fishing technology.

Recent workers have questioned why man should ever have taken up agriculture at all since, as Cohen (1977) puts it, agriculture seems to provide a poorer diet less reliably and with greater labour costs than hunting and gathering. Certainly food crops such as maize and manioc, which are efficient calorie producers eminently suitable for agricultural staples, spread much more slowly along the coast of Peru than cotton or gourds, indicating that farming was not an immediately attractive form of subsistence. Bray (1976) suggests that development of agriculture is analogous to biological evolution and that at some point during the process of gradual change a critical threshold is passed and an irreversible commitment to agriculture made. Although Bray does not believe that a single variable acted in cause-and-effect fashion to bring about change, he does consider that increase in population was probably an important factor in the evolution of more efficient ways of producing food. One advantage that Cohen (1977) concedes for agriculture is that it makes it possible to produce more food per unit space per unit time and hence to support a larger population than under a hunting and gathering regime. Cohen (1977) and Patterson (1971b) both claim that population increases towards the end of the Lithic necessitated abandoning the earlier hunting and gathering existence and concentrating instead on the more prolific resources of the ocean. Bray (1976) considers that sedentism would lead to a further population increase since ethnographic data from various sources indicate that the birth interval between children decreases when young children no longer have to be carried on nomadic migrations. Whether the comparatively short seasonal journeys of Lithic hunters and gatherers between lomas and river valleys would limit birth rates in this way is not certain, but the population increase which Cohen (1977) and Patterson (1971b) detect towards the end of the Lithic suggests that they did not.

Paulsen (1976) has suggested that dry periods in the Peruvian highlands prompted populations to move out of the seasonally arid inter-Andean valleys and down to the coast. She associates the spread of first the Huari and later the Inca Empire with such dry periods. Earlier arid phases might have led to similar emigration from the highlands, thus increasing population pressure in the coastal valleys and along the littoral. However, there is as yet no convincing evidence of such early movements in the archaeological record. The data suggest instead a well-developed network of trade routes linking coast, highlands and even the eastern jungles and involving such products as obsidian, chert, agricultural produce and, after 3000 BP, copper (MacNeish, Patterson and Browman, 1975).

Although sedentism is no longer seen as a prerequisite for development of agriculture in the way that Sauer (1952) suggested, sedentism may have had a different importance under the agriculturally difficult conditions of the Peruvian coast. Moseley (1975) has suggested that the social organisation and traditions of corporate labour manifested in the construction of the ceremonial pyramids and communal buildings of the large marine-based settlements of the Cotton Preceramic were necessary also for construction and maintenance of the elaborate canal systems, on which large-scale cultivation of land along much of the coast depended.

So far the coast has been considered as an autonomous zone, in which cultural developments took place in response to factors originating in the ecological context of the coast. However, the coast is unlikely ever to have been completely isolated from the highlands. Indeed, suggestions of early seasonal transhumance between coast and highlands and the evidence of trade already discussed show that there was considerable interaction between the two regions throughout prehistory. The river valleys provide natural and easy routes for communication between coast and highlands, giving rise to Murra's (1968) concept of 'verticality' in Andean cultural ecology. The coastal valleys, on the other hand, are separated by stretches of desert which would impede inter-valley communication. It is therefore necessary to consider how far the subsistence changes discussed for the coast were influenced by developments in the highlands.

Unfortunately comparatively few highland sites covering the period 9000-3500 BP have so far been excavated. As would be expected, agriculture seems to have begun much earlier in the highlands than on the coast. Fully domesticated common and lima beans have been found in Guitarrero Cave, certainly at 9000 BP and possibly as early as 10,500 BP (C. E. Smith, 1977). Cultivation of plants on the coast may therefore have started as a consequence of contact with the highlands. Other crops also spread to the coast from the highlands, for example chili pepper (10,500 BP in Guitarrero Cave, 4500 BP on the coast) and maize (possibly 6000 BP in Guitarrero Cave and 5000 BP in the Ayacucho caves, about 4000 BP on the coast) (C. E. Smith, 1977; Flannery, 1973).

In the highlands, as on the coast, dependence on agriculture developed only slowly. One major difference between coast and highlands is that the highlands lack a non-seasonal subsistence resource comparable to the sea which would have permitted early sedentism. Although details are not yet clear, it seems that although the highlands had domesticated plants earlier than the coast, complete dependence on agriculture may have come later in the highlands than on the coast.

While the arrival of agriculture may reflect highland influences on the coast, development of extensive irrigation systems on the coast also presupposes peaceful coexistence between coastal and up-valley populations, if not authoritarian control of the entire valley system to ensure adequate partitioning of water along its length. Whatever the nature of these intra-valley systems of water management, they must also have promoted exchange and interaction between coastal and highland peoples.

Man's adaptation to the desert coast of Peru has therefore changed through time, in response to advances in technology and social organisation. Much of this development seems to have taken place on the coast itself, elicited by the ecological and environmental conditions of the coastal region, but coast and highlands were in virtually continuous contact and the highlands were the source of at least one innovation which eventually had a great impact on coastal subsistence, namely cultivation of plants. The coastal civilisations ultimately achieved a mastery of their unique and somewhat unpredictable environment that was arguably more effective than European exploitation of this region in the centuries after the Conquest.

BIBLIOGRAPHY

Adams, R. M., 1960. Early civilisations, subsistence and environment. In C. H. Kraeling and R. M. Adams, (eds.), City Invincible, pp. 269-295. University of Chicago Press, Chicago.

Beek, K. J. and Bramao, D. L., 1968. Nature and geography of South American soils. In E. J. Fittkau, J. Illies, H. Klinge, G. H. Schwabe and H. Sioli, (eds.), Biogeography and Ecology in South America vol. 1, pp. 82-112. Monographiae Biologicae 18. Dr. W. Junk N.V., The Hague.

Bowman, Isaiah, 1916. The Andes of Southern Peru. 1st edition. Henry Holt & Co., New York.

Bray, Warwick, 1976. From predation to production: the nature of agricultural evolution in Mexico and Peru. In G. de G. Sieveking, I. H. Longworth and K. E. Wilson, (eds.), Problems in Economic and Social Archaeology, pp. 73-95. Duckworth, London.

Bryan, A. L., 1973. Palaeoenvironments and cultural diversity in Late Pleistocene South America. Quaternary Research 3, 237-256.

CLIMAP. 1976. The surface of the Ice Age earth. Science 191, 1131-1137.

Cohen, Mark N., 1977. Population pressure and the origins of agriculture: an archaeological example from the coast of Peru. In Charles A. Reed, (ed.), Origins of Agriculture, pp. 135-177. Mouton Publishers, The Hague.

Colinvaux, P. A., 1972. Climate and the Galapagos Islands. Nature 240, 17-20.

Craig, Alan K. and Psuty, Norbert P., 1971. Palaeoecology of shell mounds at Otuma, Peru. Geog. Rev. 61, 125-132.

Cutler, Hugh C. and Whitaker, Thomas W., 1969. A new species of Cucurbita from Ecuador. Ann. Missouri Bot. Gard. 55, 392-396.

D'Antoni, H. L. and Solbrig, O. T., 1977. Algarrobos in South American cultures past and present. In B. B. Simpson, (ed.), Mesquite: Its Biology in Two Desert Scrub Ecosystems, pp. 189-199. US/IBP Synthesis Series vol. 4, Dowden, Hutchinson & Ross Inc., Stroudsburg, Pennsylvania.

Drosdoff, M., Quevedo, F. and Zamora, C., 1960. Soils of Peru. 7th Int. Congr. Soil Sci. 13, 97-104.

Eidt, R. C., 1968. The climatology of South America. In E. J. Fittkau, J. Illies, H. Klinge, G. H. Schwabe and H. Sioli, (eds.), Biogeography and Ecology in South America vol. 1, pp. 54-81. Monographiae Biologicae 18. Dr. W. Junk N.V., The Hague.

Engel, Frederic A., 1970. Exploration of the Chilca Canyon, Peru. Current Anthropology 11, 55-58.

Engel, Frederic A., 1973. New facts about pre-Columbian life in the Andean lomas. Current Anthropology 14, 271-280.

Engel, Frederic A., 1976. An Ancient World Preserved: Relics and Records of Prehistory in the Andes. Translated by Rachel Kendall Gordon. Crown Publishers Inc., New York.

Farrington, Ian , 1974. Irrigation and settlement pattern: preliminary research results from the north coast of Peru. In Theodore E. Downing and McGuire Gibson, (eds.), Irrigation's Impact on Society, pp. 83-94. University of Arizona Press, Tucson, Arizona.

Fenner, C. N., 1948. Pleistocene climate and topography of the Arequipa region, Peru. Bull. Geol. Soc. Am. 59, 895-917.

Flannery, K. V., 1973. The origins of agriculture. Ann. Rev. Anthrop. 2, 271-310.

Flannery, K. V., Kirkby, A. V. T., Kirkby, M. J. and Williams, A. W., 1967. Farming systems and political growth in ancient Oaxaca. Science 158, 445-454.

Fuenzalida V., Humberto, Cooke, Donald, Paskoff, Roland, Segerstrom, Kenneth and Weischet, Wolfgang, 1965. High stands of Quaternary sea level along the Chilean coast. In H. E. Wright, Jr. and David G. Grey, (eds.), International Studies on the Quaternary, pp. 473-496. Special Paper 84, Geological Society of America, New York.

Garner, H. F., 1959. Stratigraphic-sedimentary significance of contemporary climate and relief in four regions of the Andes mountains. Bull. Geol. Soc. Am. 70, 1327-1368.

Gates, W. L., 1976. Modelling the Ice Age climate. Science 191, 1138-1144.

Grobman, Alexander and Bonavía, Duccio, 1978. Pre-ceramic maize on the north-central coast of Peru. Nature 276, 386-387.

Hastenrath, S., 1971. On snowline depression and atmospheric circulation in the tropical Americas during the Pleistocene. S. Afr. Geog. J. 53, 53-69.

Hawkes, J. G., 1967. The history of the potato. Journ. Roy. Hort. Soc. 92, 207-224.

Idyll, C. P., 1975. The anchovy crisis. Scientific American 228, 22-29.

Kautz, R. R. and Keatinge, R. W., 1977. Determining site function: a north Peruvian coastal example. Amer. Antiq. 42, 86-97.

Lanning, Edward P., 1963. A pre-agricultural occupation on the central coast of Peru. Amer. Antiq. 28, 360-371.

Lanning, Edward P., 1965. Early man in Peru. Scientific American 213, 68-76.

Lanning, Edward P., 1967. Peru Before the Incas. Prentice-Hall Inc., Englewood Cliffs, New Hersey.

Lanning, Edward P. and Patterson, Thomas C., 1967. Early man in South America. Scientific American 217, 44-50.

León, Jorge, 1964. Plantas Alimenticias Andinas. Boletin technico 6, Instituto Interamericano de Ciencias Agricolas Zona Andina. Lima, Peru.

Llagostera Martinez, Agustin, 1979. 9,700 years of maritime subsistence on the Pacific: an analysis by means of bioindicators in the north of Chile. Amer. Antiq. 44, 309-324.

Lynch, Thomas F., 1974. The antiquity of man in South America. Quaternary Research 4, 356-377.

MacNeish, Richard S., 1977. The beginnings of agriculture in central Peru. In Charles A. Reed, (ed.), Origins of Agriculture, pp. 753-801. Mouton Publishers, The Hague.

MacNeish, Richard S., Patterson,Thomas C. and Browman,David L., 1975. The Central Peruvian prehistoric interaction sphere. Papers of the Robert S. Peabody Foundation for Archaeology 7, Andover, Massachusetts.

Manabe, S. and Hahn, D. G., 1977. Simulation of the tropical climate of an Ice Age. J. Geophys Res. 82, 3889-3911.

Martins, Rena, 1976. New archaeological techniques for the study of ancient root crops in Peru. Ph.D. thesis, University of Birmingham.

Matsuzawa, Tsugio, 1978. The Formative site of Las Haldas, Peru: architecture, chronology and economy. Amer. Antiq. 43, 652-673.

Meigs, Peveril, 1966. Geography of Coastal Deserts. Arid Zone Research 28, UNESCO, Paris.

Moseley, M. Edward, 1969. Assessing the archaeological significance of mahamaes. Amer. Antiq. 34, 485-487.

Moseley, M. Edward, 1974. Organisational preadaptation to irrigation: the evolution of early water-management systems in coastal Peru. In Theodore E. Downing and McGuire Gibson, (eds.), Irrigation's Impact on Society, pp. 77-82. University of Arizona Press, Tucson, Arizona.

Moseley, M. Edward, 1975. The Maritime Foundations of Andean Civilisation. Cummings, California.

Moseley, M. Edward and Mackey, Carol J., 1972. Peruvian settlement pattern studies and small site methodology. Amer. Antiq. 37, 67-81.

Murra, John V., 1968. An Aymara Kingdom in 1567. Ethnohistory 15, 115-151.

Parmalee, Paul W. and Klippel,Walter E., 1974. Freshwater mussels as a prehistoric food resource. Amer. Antiq. 39, 421-434.

Parsons, Jeffrey R., 1968. The archaeological significance of mahamaes cultivation on the coast of Peru. Amer. Antiq. 33, 80-85.

Parsons, Jeffrey R. and Psuty, Norbert P., 1975. Sunken fields and prehispanic subsistence on the Peruvian coast. Amer. Antiq. 40, 259-282.

Parsons, Mary Hrones, 1970. Preceramic subsistence on the Peruvian coast. Amer. Antiq. 35, 292-304.

Patterson, Thomas C., 1971. The emergence of food production in central Peru. In Stuart Struever, (ed.), Prehistoric Agriculture, pp. 181-207. Natural History Press, Garden City, New York.

Patterson, Thomas C., 1971. Central Peru: its population and economy. Archaeology 24, 316-321.

Paulsen, Allison C., 1976. Environment and empire: climatic factors in prehistoric Andean culture change. World Archaeology 8, 121-132.

Pickersgill, Barbara, 1969. The archaeological record of chili peppers (Capsicum spp.) and the sequence of plant domestication in Peru. Amer. Antiq. 34, 54-61.

Quinn, W. H., 1971. Late Quaternary meteorological and oceanographic developments in the equatorial Pacific. Nature 229, 330-331.

Reparaz, G. de, 1958. La zone aride du Perou. Geografiska Annaler 40E, 1-62.

Richards, Horace G. and Broecker, Wallace, 1963. Emerged Holocene South American shorelines. Science 141, 1044-1045.

Richardson, James B., III, 1973. The preceramic sequence and the Pleistocene and post-Pleistocene climate of northwest Peru. In D. W. Lathrap and J. Douglas, (eds.), Variation in Anthropology: Essays in Honour of J. C. McGregor, pp. 199-211. Illinois Archaeological Survey, Urbana, Illinois.

Rostworowski de Diez Canseco, Maria, 1977. Coastal fishermen, merchants and artisans in pre-hispanic Peru. In Elizabeth P. Benson, (ed.)., The Sea in the Pre-Columbian World, pp. 167-188. Dumbarton Oaks Research Library Collection, Washington, D.C.

Rowe, J. H., 1948. The Kingdom of Chimor. Acta Americana 4, 26-59.

Rowe, J. H., 1969. The sunken gardens of the Peruvian coast. Amer. Antiq. 34, 320-325.

Sarma, Akkaraju V. N., 1974. Holocene palaeoecology of south coastal Ecuador. Proc. Am. Phil. Soc. 118, 93-134.

Sauer, Carl O., 1952. Agricultural Origins and Dispersals. American Geographical Society, New York.

Simpson, Beryl B., 1975. Glacial climates in the eastern tropical South Pacific. Nature 253, 34-36.

Smith, C. Earle, Jr., 1977. Recent evidence in support of the tropical origin of New World crops. In David S. Seigler, (ed.), Crop Resources, pp. 79-95, Academic Press, New York.

Smith, R. T., 1979. The development and role of sunken field agriculture on the Peruvian coast. Geog. J. 145: 387-400.

Stephens, S. G. and Moseley Edward M., 1974. Early domesticated cottons from archaeological sites in central coastal Peru. Amer. Antiq. 39, 109-122.

Towle, Margaret A., 1961. The Ethnobotany of Pre-Columbian Peru. Viking Fund Publications in Anthropology 30. Aldine Publishing Co. Chicago.

Tricart, J., 1963. Oscillations et modifications de caractère de la zone aride en Afrique et en Amérique Latine lors des periodes glaciaires des hautes latitudes. In Changes of Climate: Proceedings of the Rome Symposium organised by UNESCO and the World Meteorological Organisation. Arid Zone Research 20, 415-419. UNESCO, Paris.

van der Hammen, T., 1974. The Pleistocene changes of vegetation and climate in tropical South America. J. Biogeogr. 1, 3-26.

West, Michael, 1979. Early watertable farming on the north coast of Peru. Amer. Antiq. 44, 138-144.

Whitaker, T. W. and Davis, G. N., 1962. Cucurbits. World Crop Books, Leonard Hill Ltd., London.

Williams, M. A. J., 1975. Late Pleistocene tropical aridity: synchronous in both hemispheres? Nature 253, 617-618.

Zavaleta, A. G., 1965. The nature of saline and alkaline soils of the Peruvian coastal zone. Agrokémia és Talajtan 14 suppl., 415-421.

SEAWEED AS A PREHISTORIC RESOURCE

By Martin Bell

Though many of the modern foods, fodders and fertilizers are based on post-medieval inventions or are introductions from overseas, I think it fair to say that field archaeologists have not always been sufficiently on the look out for evidence of native resources which might formerly have been more extensively exploited. Seaweeds are a case in point, the ethno-historical sources show that they have been an important resource in a number of coastal areas. Can we hold out any hope of finding evidence for their exploitation on archaeological sites? The marine algae are after all fleshy, generally non-skeletal and lacking in specialised strengthening tissue and might be expected not to survive under any but the most exceptional preservation conditions.

We shall be concerned with three groups: Chlorophyta, green algae; Phaeophyta, brown algae; and Rhodophyta, red algae. Some members of the latter are distinguished by a calcified thallus and evidence of them may therefore survive under calcareous conditions.

Ethno-Historical Evidence.

Communities exploiting seaweeds may collect the cast or drift weed, which is generally thrown up by storms, or they may cut the weed from the rocks either at low tide or by diving. Under exceptional circumstances, where the marine substrate was not capable of satisfying demand, seaweeds were cultivated. This has been done in Tokyo Bay since about 1700 A.D. (Chapman 1950, p. 163) and in south west Ireland where boulders were laid down in the sand in neat rows to encourage the growth of *Fucus* and *Ascophyllum* (Cotton 1912, p. 153); how far this practice goes back is unknown.

Many seaweeds possess interesting biochemical properties which are the basis of much of their economic importance, for instance they are capable of concentrating salts of potash and soda from the sea's dilute solution. Iodine and potassium in particular are concentrated and species of *Laminaria* are capable of storing up to 30,000 times the iodine content of ocean water (Levring *et al*. 1969, p. 347). During the post-medieval period brown rockweeds were burnt to produce kelp; this is first documented in the seventeenth century when the French were extracting soda for glazing and glass. Kelp began to be produced in Britain around 1720, the industry was based upon areas where large quantities of weed were thrown up by storms, chiefly Scotland, the Northern Isles and western Ireland. In some areas the weed was dried on specially built stone walls and burnt in circular kilns or trenches. One such has been excavated on North Ronaldsay (Fenton 1978), and examples exposed by erosion have recently been investigated in Birsay Bay on Orkney (J. Rackham, pers. comm.). These kilns produced a solid cake of kelp which was then sent to the chemical factory. Originally production was for soda, but from 1814 it was also a major source of iodine until the discovery of mineral deposits in Chile in 1873 (Chapman 1950).

There is a little evidence of much earlier industrial and craft usage of seaweeds. Pliny, in his Natural History, mentions garments being dyed purple with seaweed extract and during more recent times in Wales red dyes have been obtained from the same source (Stephenson 1968, p. 42). They also seem to have been used for medical purposes from a very early date; a Chinese herbal of 2700 B.C. refers to seaweeds (Levring et al. 1969), and they were a common folk remedy for a variety of complaints in Ireland. Other recorded minor uses are for glue, caulking boats and occasionally even the production of artifacts such as containers, fishing lines, hooks, etc.

None of these multifarious uses, interesting as they are, is likely to have been a major reason for the exploitation of seaweeds in prehistory when they would have been much more important as food and fertilizer. Their earliest recorded use as animal feed is by Linnaeus who, in 1745, records the mixing of boiled bladderwrack with bran to produce a pigfeed, on North Ronaldsay and blackfaced sheep feed entirely on seaweed for ten months on the year, and are only pastured during the lambing season (Round 1973, p. 221). The use of this fodder was not just restricted to coastal pastures, and even when animals were taken up to the shielings it was the practice to carry a few basket loads of seaweed on the back of a donkey; on Lewis this was stored in a special little hut, the tigh-an-fhaomainn, or seaweed house (Fenton 1976, p. 135).

Seaweeds have also been extensively exploited for human consumption particularly in Japan and the Far East. Aboriginal populations in Austrialia and the Maoris of New Zealand regularly used seaweeds, such as Porphyra, in salads and soups. Their earliest recorded usage in the west comes from an Icelandic saga of 961 A.D., which refers to the edibility of sol (Rhodymenia palmata). The right to collect sol is recorded in Icelandic lawbooks as early as the first half of the twelfth century by which time, according to Hallsson (1964, p. 398), it seems to have been collected and transported over long distances. One suspects an equally ancient tradition of seaweed eating existed in the coastal areas to the west and north of the British Isles where Porphyra is still eaten as a constituent of laver bread.

Prior to the industrial revolution undoubtedly the most important use of seaweed was as fertilizer, this use was first mentioned by Palladius writing in the fourth century A.D., and it was being utilised in France in the twelfth century A.D. Camden (1772, p. 158) records its use in Devon in 1586. Clearly its beneficial effects upon the soil properties were appreciated from an early date, for instance Falle (1734, p. 149), writing about Jersey in 1694 describes how seaweed 'ameliorates and fertilizes the earth imbibing itself into it, softening the clod and keeping the root of the corn moist during the most parching heats of summer'. In fact the utilisation of seaweeds is an important factor in the formation of very fertile, partly man-made, soils which stretch up the North Atlantic littoral from Portugal to the Northern Isles. In Portugal there are market gardens known as maceiras, which are believed to be of some antiquity and which are intenstively cultivated by hand for early vegetables using seaweed as fertilizer (Larkcom 1977, 1979). In North West France there is a 400 mile long strip of coast known as the ceinture doré, where the steady use of seaweed on a strip 500 m from the shore has produced a soil which is capable of the continuous production of best quality vegetables without any rotation (Chapman 1950, p. 134).

Similar man-made soil phenomena are the lazy-beds of Ireland and Scotland. These were formed by laying manure, generally seaweed, in parallel strips about 1 m wide and the same distance apart. Turf and earth from between the strips was then spade-dug and laid on top of the fertilizer to produce cultivation ridges. Latterly lazy-beds have been used mainly for potato growing, but their use for growing grain crops is documented from at least the early sixteenth century (Fenton 1976). Furthermore recent excavations at Belderg, Co. Mayo, have produced evidence of cultivation ridges, of lazy-bed type, dating between 4500 and 3200 B.P. (Mitchell 1976, p. 141). In Ireland marine algae have also played a part in the formation of plaggen soils which have a deep humic profile produced by the importation of material, including shell, sand or coralline algae. The latter are calcareous members of the Rhodophyta, the broken fragments of which make up the dazzling coral beaches of Connemara. This calcareous sand was collected and spread on peaty soils in south west Ireland (Cotton 1912, p. 153) often mixed with byre manure and seaweed. Conry and Mitchell (1971) doubt whether the true deep plaggen soils of Ireland formed much before the land hunger of the seventeenth century, but the agricultural practices which produced them are undoubtedly of much more ancient origin, there are references to the use of calcareous sand in pre-Christian Ireland and the use of coralline algae is documented in Normandy from the twelfth century.

Preservation

This wealth of ethno-historical evidence establishes a reasonable *a priori* case for the exploitation of seaweeds in prehistory, but is there any way in which we may obtain evidence of this? Seaweed itself is only likely to be preserved under exceptional conditions. The geological record is almost entirely restricted to calcareous algae although occasionally fleshy parts are represented by carbonaceous films and imprints (Wray 1977). At one exposure in Canada, where burial appears to have been very sudden, there were well preserved blades of *Laminaria* and other species dating from 10,800 ± 150 radiocarbon years b.p. (Illman *et al*. 1970). As far as preservation on archaeological sites is concerned, the evidence is very limited. The red algae *Gigartina chauvinii* was identified from pre-Columbian deposits in Peru dating to the fourteenth century A.D., and interestingly enough this is a species which is still eaten in Peru today (Dixon 1973, p. 230). Burnt seaweed has been recovered from the Brough of Birsay on Orkney and, although the material was very fragmentary, some of it had characteristic features of *Fucus vesiculosus* (Alison Donaldson, pers. comm.). The site is one with both Norse and pre-Norse phases but at present there is no closer dating evidence for the context which produced this material. Another find of carbonised material, tentatively interpreted as burnt seaweed, is reported from recent excavations at 'Da Biggins', Shetland, a Norse site of the twelfth to thirteenth centuries (Crawford 1979).

We should not be surprised by the paucity of preserved non-calcareous algae since one of their virtues as a fertilizer is the speed with which they break down. Experimental work on large brown seaweeds like *Macrocystis*, *Ecklonia* and *Durvillea* shows that they decompose within four months; *Carpophyllum*, on the other hand, was still recognisable after a year (Chapman 1950, p. 253). Unfortunately very little work seems to have been done on decay products or residuals which might be traced microscopically or chemically.

Date and where collected		26.1.79 Cast 3kg.			4.2.79 break-water	4.3.79 break-water	23.5.79 cast ·27kg.		23.5.79 break-water		
Weight or percentage collected		28%	70%	2%	·33kg	·20kg	80%	20%	·24kg		4·04 kg.
Inclusions	Seaweed etc. species	Laminaria saccharina (L.)	Fucus vesiculosus (L.)	Flustra foliacea (L.)	Fucus spiralis (L.)	Fucus spiralis (L.)	Laminaria saccharina (L.)	Laminaria digitata (Huds)	Fucus spiralis (L.)		
Bivalve Molluscs	Mytilus edulis L. [adult, frag.]	8	5		1	1	8	1			24
	? Mytilus edulis L. [juvenile]	11					27	2			40
	? Modiolus modiolus (L.) [juvenile]						3				3
	? Venerupis sp. [juvenile]						1				1
	Sphenia binghami Turton	1									1
	Pholas dactylus L. [frag.]	3	1				6				10
	Unidentified bivalve	1									1
	Unidentified bivalve [frag.]	37	10	3		3	22	3			78
Gastropods	Patella vulgata L. [frag.]	1					1				2
	? Littorina littorea (L.) [frag.]						1				1
	Littorina saxatilis (Olivi)				*				1		1
	Rissoa parva (da Costa)	1					2				3
	Crepidula fornicata (L.) [frag.]	1									1
	Unidentified gastropod [frag.]	1									1
Barnacles	Mainly Chthamalus sp. [adult, mainly frags.]	71			26	13	28	23	3		164
	Juvenile barnacles						1	5	2		8
Bryzoans	Membranipora membranacea (L.)	4					5	1			10
Forams	Massilina secans (d'Orbigny)						7				7
Polychaete Worms	Family ? SABELLARIIDAE [frags.]	3					9	5			17
	? Hydroides norvegica Gunnerus						1	11			12
	unknown serpulid [frag.]	3									3
	Spirorbis spirorbis L.		119								119
Gravel grade shingle		99	104	1	6	2			3		215

* Present in large numbers feeding on algae between the strands of F. spiralis.

Table 1. Inclusions among seaweeds collected on the shore at Saltdean, Sussex.

Inclusions with Modern Seaweeds

If seaweed itself is seldom preserved then perhaps there is some potential in demonstrating its utilization by less direct methods, for instance the presence of seaweed epifaunas or other material which might have been transported to sites with seaweed. In order to assess the potential of this type of evidence I collected *Laminaria saccharina*, *Fucus vesiculosus* and *Laminaria digitata* as cast weed from the beach at Saltdean, Sussex, together with *Fucus spiralis* which was growing on a nearby breakwater. The holdfasts were dissected and the fronds and stipes were carefully examined for any materials which were likely to be preserved in an archaeological context; the results are presented in Table 1. The most productive were the holdfasts of cast *Laminaria* sp. Bivalve molluscs included tiny juveiles of ? *Mytilus edulis*, ?*Modiolus modiolus* and ?*Venerupis* sp. together with examples of *Sphenia binghami* and fragments of the common piddock, *Pholas dactylus.* The gastropods included a number of small species, *Littorina saxatilis*, *Rissoa parva* and *Crepidula fornicata*. Particularly interesting was the discovery of a large number of barnacles within the holdfasts. The lace-like bryozoan *Membranipora membranacea*, which has calcified side walls, was found growing on the stipes of *Laminaria* sp. Seven forams, identified as *Massilina secans*, were found on *Laminaria saccharina* and, since these were not specifically searched for, smaller forams are likely to have been missed during initial sorting. Trapped within the holdfasts were also the sandy and calcareous tubes of polychaete worms, and on the fronds of *Fucus vesiculosus* were found 119 examples of a species of spirorbid which generally grows on *Fucus* or stones. Finally there were a large number of pieces of gravel-grade shingle trapped within the holdfasts. Obviously this is but a selection of the things which might have been introduced to sites with seaweed, although it must also be emphasised that virtually all of this material could have arrived on sites in a variety of other ways, for instance on stones, driftwood, etc. This survey does however, provide a background against which to review some pieces of evidence from archaeological sites.

Archaeological evidence

Not surprisingly some of the most interesting archaeological evidence comes from the Atlantic seaboard where seaweed utilisation is so well attested historically. Fowler and Thomas (1962) excavated ridge-and-furrow fields at Gwithian, Cornwall, dating between the ninth century and *c.* 1100 A.D. They found a number of very tiny shells of mussel and periwinkle which were far too small for foodstuff, and also various little beach pebbles, all of which seems to point to the use of seaweed as manure. This is particularly interesting in view of the indications of systematic manuring, in this case not necessarily with seaweed, from earlier Bronze Age contexts at Gwithian (Megaw *et al.* 1960). At the Pictish and Viking age farmstead at Buckquoy, Orkney (Ritchie 1977; p. 251) seashore resources studied by Dr. John Evans and Miss Penny Spencer include a number of items which may have been brought to the site with seaweed: these include *Patina pellucida*, which lives on *Laminaria*; ?*Modiolus modiolus*; *Littorina littoralis*; *Spirorbis borealis* (now *Spirorobis spirorbis*); and *Flustrella hispida*, a bryozoan which forms encrusting colonies on *Ascophyllum* and *Fucus* spp. An important caveat should be added, however, because a few spirorbids and ?Flustrella sp. were found adhering to marine molluscs. A distinct horizon containing Littorina littoralis is also reported in coastal blown sand deposits at Freshwater West in Dyfed and may point to the collection of seaweeds (Evans 1979).

There is also evidence of seaweed exploitation in areas where it is very little attested in historical times. A case in point is the multi-period settlement at Bishopstone on the Sussex coast (Bell 1977). Within the settlement a number of examples of inedible or very small marine molluscs seemed likely to have been introduced with seaweed. These included examples of Anomia ephippium, which commonly attaches itself to holdfasts or large seaweeds (Yonge 1949, p. 166); Littorina saxatilis, which feeds on and around seaweeds on the higher shore, and thirteen examples of Littorina littoralis, which is generally found on Fucus and Ascophyllum (Graham 1971). These molluscs were in contexts dating from the early Iron Age to the pagan Saxon period. Another interesting pointer was the presence of barnacles in two pits; four pieces in early Iron Age pit 423 and 108 pieces in Romano-British pit 917. This evidence was considerably strengthened by the presence of four bryozoans which were identified as Turbicellepora avicularis, one of which was unstratified, two came from an early Iron Age ditch and one was found on the floor of Middle Iron Age pit 791. This species is fully marine and characteristic of deepish water below about 20m. They are liable to be washed up after a storm and Miss Pat Cook of the British Museum (Natural History) tells me that she has found them among seaweed collected for manure. Attention should also be drawn to the existence in the Bishopstone settlement of large quantities of beach shingle, pebbles and pieces of chalk exhibiting the borings of marine molluscs. Both could have been deliberately brought to the site but they might also have been accidentally introduced with seaweed. These various strands of evidence certainly suggest that seaweeds were being exploited by the Bishopstone settlement. Furthermore the discovery of some of these things in Iron Age and Romano-British pits makes one wonder whether they were storing seaweeds, for whatever purpose. Certainly this was done in Iceland where stores were laid up in trenches where they were compressed with heavy oak planks and stones (Chapman 1950, p. 124).

Evidence from the Bishopstone settlement was complemented by Dr. Ken Thomas' study of molluscan evidence from a nearby lynchet (Bell 1977, p. 258). Among the terrestrial fauna there were a few aquatic molluscs: Mytilus edulis, Littorina littorea, Littorina saxatilis, the estuarine species Hydrobia ulvae and freshwater ?Unio sp. The marine molluscs among these support the evidence from the settlement and suggest that seaweed was being used as manure.

It is unfortunate that very few excavators seem to have attempted to make total collections of marine molluscs; one place where this has recently been done is the Bullock Down multi-period settlement project on the coast at Beachy Head, East Sussex (Caroline Cartwright, pers. comm.). As on most sites the vast majority of the molluscs were edible species but here again we find small numbers of Littorina littoralis turning up within the structures and ploughsoil of the Kiln Combe Medieval settlement as well as one or two other species which are likely to have come in as 'dead' shells.

There is also the possibility of identifying distinctive insect faunas which accompany rotting heaps of seaweed on the beach. The insects from the Roman well at Rudston villa, Yorkshire, some 10 km from the sea (Buckland 1980) included a small spider *Erigone? artica* which is generally found along the strandline among seaweed. Although the members of this genus are active aeronauts, and could have reached the well naturally, there is some support for the hypothesis, from the presence of beach pebbles in the well, that the villa's inhabitants were collecting seaweed. Comparable evidence derives from Dr. Peter Osbourne's work (pers. comm.).on insect faunas from the Guildhall site in Exeter. Material from a sixteenth century pit included large numbers of puparia of the fly *Leptocara zosterae* which, in its larval state, is generally found among rotting seaweed. The associated beetle fauna is one which would be at home in a wet compost heap containing dung and household refuse.

Another interesting pointer is the presence in Ipswichian contexts at Bacon Hole, Gower, of Foraminifera which are indicative of a shallow water littoral environments and are gernally associated with seaweeds (Stringer 1977, p. 38). In the particular case of this coastal cave there is no suggestion that the seaweed was being exploited but it does emphasise the potential importance of forams for future studies.

We should also bear in mind the possibility of early industrial utilisation of seaweed. Kelp has, for instance, been suggested as a possible source of alkali in the production of faience beads (Newton and Renfrew 1970). The type of evidence which might be sought for this on archaeological sites is suggested by burnt shells of *Littorina littoralis* and *Patina pellucida* from an eighteenth century site on the Isles of Scilly, for which there is also archaeological evidence of kelping (Evans 1978, p. 33). It might also be worth investigating the non-metallurgical slags which result from the vitrification of wood ash, seaweed ash, cow dung, burnt straw, etc. Slags of this kind were present at Bishopstone (Bell 1977, p. 289) and on the Essex red hills (Petrie 1910; Jenkins 1910), and the indications are that analytical studies might be able to establish from which materials these slags derived (Evans and Tylecote 1967). The burning of seaweed might, of course, have been undertaken for a variety of purposes apart from potash extraction; dried seaweeds were used as fuel in Iceland (Hallsson 1964) and Scotland (Fenton 1978, p. 206), and on the Channel Islands seaweed was sometimes ashed before being used as fertilizer (Stephenson 1968, p. 47).

Conclusions

In conclusion it cannot be emphasised too strongly that virtually all the materials which I have suggested were introduced to sites with seaweed might have found their way there by a variety of other means. However, the evidence becomes more convincing when, as at Buckquoy and Bishopstone, we have a number of varieties of evidence all contributing to the same conclusion. At Bishopstone and Gwithian we do seem to be on fairly safe ground in postulating the use of seaweed as manure. Seen in the context of evidence for prehistoric marling, the spreading of domestic and farmyard manure, coralline algae, shell sand and shells from the midden, this suggests that prehistoric communities husbanded their pedological resources rather more carefully than we have sometimes tended to give them credit for. The very longevity of many settlements and the extent of landscape utilisation, now being revealed by field surveys, would have necessitated the intensive manuring of an infield. Under certain circumstances, particularly where the natural soil was very poor, these measures might ultimately have led to the creation of a largely man-made soil such as the plaggen soils and lazy-beds.

The origins of these soils and the extent of prehistoric manuring in general are matters of the greatest interest and this paper has suggested some lines for future enquiry. Few of the biological remains recorded as inclusions in the modern seaweed study are likely to be spotted during excavations. Further investigations of field systems near the coast, particularly calcareous deposits, should employ sieving programmes specifically to search for the smaller molluscs, spirorbids and barnacles; investigations should also be made for forams. The same methods should be extended to settlement contexts where larger molluscs or other biological remains suggest the presence of seaweeds.

Acknowledgements

I am grateful to the following members of the staff of the British Museum (Natural History) for identifying material or checking my identifications: Pat Cook, Kathie Way, Solene Whybrow, Dr. Boxall and Alex Muir. The following people have allowed me to refer to their unpublished work: Dr. Paul Buckland, Caroline Cartwright, Alison Donaldson, Dr. Peter Osbourne and James Rackham. I have also had the benefit of discussion on this subject with Terry O'Connor, Peter Boyd and particularly Dr. Ken Thomas who kindly read an earlier version of this paper and suggested a number of improvements.

BIBLIOGRAPHY

Bell, M., 1977. 'Excavations at Bishopstone', Sussex Archaeo. Colls. 115: 1-299.

Buckland, P. C., 1980. 'Insect remains from the well' In I. M. Stead, Rudston Roman villa, Yorkshire Philosophical Society, Yorks.

Camden, W., 1772. Britannia vol. 1, E. Gibson's edition, London.

Chapman, W. J., 1950. Seaweeds and Their Uses, Methuen, London.

Conry, M. J. and Mitchell, G. F., 1971. 'The age of Irish plaggen soils' In D. H. Yaalon (ed.), Paleopedology, International Society Soil Science, 129-138. Jerusalem.

Cotton, A. D., 1912. 'Marine algae', Proc. Roy. Irish Acad., 31: 1-178.

Crawford, B., 1979. 'A progress report on excavations at "Da Biggins" Papa Stour, Shetland 1978' Northern Studies, No. 13, 37-41.

Dixon, P. S., 1973. Biology of the Rhodophyta, Oliver and Boyd, Edinburgh.

Evans, J. G., 1978. An Introduction to Environmental Archaeology, Elek, London.

Evans, J. G., 1979. 'The palaeo-environment of coastal blown-sand deposits in western and northern Britain, Scottish Archaeological Forum, 9: 16-26.

Evans, R. T. and Tylecote, R. F., 1967. 'Some vitrified products of non-metallurgical significance', Bull. Hist. Metallurgy Gp., 1: 22-3.

Falle, P., 1734. Caesarea: or an Account of the Isle of Jersey, second edition, Wotton, London.

Fenton, A., 1976. Scottish Country Life, Donald, Edinburgh.

Fenton, A., 1978. The Northern Isles: Orkney and Shetland, Donald, Edinburgh.

Fowler, P. J. and Thomas, A. C., 1962. 'Arable fields of the pre-Norman period at Gwithian', Cornish archaeology No. 1, 61-84.

Graham, A., 1971. British Prosobranchs, Synopsis of the British fauna (N.S.) No. 2, Academic Press, London.

Hallsson, S. V., 1964. 'The use of seaweeds in Iceland'. In Davy de Virville and J. Feldmann (Eds.), Comptes rendus du IV congrès international des algues marines Biarritz 1961, Pergamon, New York.

Illman, W. I., et al., 1970. 'Marine algae of the Champlain Sea episode near Ottawa', Canad. J. Earth Sci., 7: 1583-1585.

Jenkins, J. H., 1910. 'Remarks on Dr. Flinders Petrie's theory', Proc. Soc. Antiq. Lond., 23: 90-96.

Larkcom, J., 1977. 'Portuguese vegetable growing', The Garden, 102, pt. 7: 289-294.

Larkcom, J., 1979. 'Random Harvest', The Observer 20.5.79, 40.

Levring, T. et al., 1969. Marine Algae: a Survey of Research and Utilisation, Cram and de Gruyter, Hamburg.

Megaw, J. V. S., et al. 1960. 'The Bronze Age settlement of Gwithian, Cornwall', Proceedings West Cornwall Field Club, new series 2: 200-215.

Mitchell, F., 1976. The Irish Landscape, Collins, London.

Newton, R. G. and Renfrew, C., 1970. "British faience beads reconsidered", Antiquity, 44: 199-206.

Petrie, W. F., 1910. 'Suggested origin of the red hills', Proc. Soc. Antiq. Lond., second series, 23: 88-90.

Ritchie, A., 1977. 'Excavation of Pictish and Viking-Age Farmsteads at Buckquoy, Orkney', Proc. Soc. Antiq. Scot., 108: 174-227.

Round, F. E., 1973. The Biology of the Algae, Arnold, London.

Stephenson, W. A., 1968. Seaweed in Agriculture and Horticulture, Faber, London.

Stringer, C., 1977. 'Evidence of climatic change and human occupation during the last interglacial at Bacon Hole Cave, Gower', Gower 28: 39.

Wray, J. L., 1977. Calcareous Algae, Elsevier, Amsterdam.

Yonge, C. M., 1949. The Sea Shore, Collins, London.

VEGETATIONAL HISTORY OF THE ISLES OF SCILLY

G. W. Dimbleby, J. R. A. Greig and R. G. Scaife

INTRODUCTION

Exposed off-shore islands such as the Isles of Scilly have for long been the subject of different views regarding the nature of the climax vegetational cover that they carried. Today these isles are devoid of indigenous forest trees; though trees are present they are all introduced. However, the fact that they do grow there today indicates that tree growth is possible, which is at variance with an alternative view that the exposure factor is so great in such situations that tree growth is physiologically impossible. On this theory some form of heathland is usually seen as the vegetational climax (Gimingham, 1972; Coombe and Frost, 1956).

The first direct evidence that deciduous forest cover once existed here was obtained by pollen analysis of the palaeosol beneath a sand-dune at Innisidgen, on the North coast of St. Mary's, the largest island in the group (Dimbleby, 1977). This analysis also provided evidence of early (but undated) agricultural practice similar to that commonly found on the mainland of Britian.

The present paper adds to the evidence on both these points. The pollen analysis of a peat deposit at Higher Moors on St. Mary's gives proof of the existence of primary mixed oak forest and subsequent clearance (RGS), whilst pollen analyses of soils associated with the Bronze Age on Nornour (JRAG) and the Iron Age at Halangy Down on St. Mary's (GWD) contribute evidence about contemporary landscapes and the role of prehistoric agriculture in them. The positions of these sites are shown on the map (Fig. 1).

PORTH HELLICK - HIGHER MOORS (RGS)

Porth Hellick Bog/Higher Moors nature reserve is one of two areas of mire peat occurring on St. Mary's. A second site, not discussed here, is at Lower Moors (SV913106), one kilometre east of Hugh Town. Both have been sampled for pollen analyses to help elucidate the vegetational history of the Isles of Scilly and to provide a framework within which soil pollen work can be correlated.

Higher Moors is a topogenous mire extending inland (NNW) from Porth Hellick Bay (SV925107) to Holy Vale (SV921115). There is substantial evidence that peat from both mires has been used for fuel in the historical past as peat cuts and baulks can be seen. The effect of this on the vegetation status has been the drying out of the mire and subsequent scrub invasion. More recently, continued land drainage and extraction of ground water for public supply has lowered the water table further, causing degradation of the mire. The area of Higher Moors nature reserve consequently consists of a mosaic of wetland vegetation communities ranging from <u>Salix</u> (willow) (<u>S</u>. <u>caprea</u>, <u>S</u>. <u>viminalis</u>)

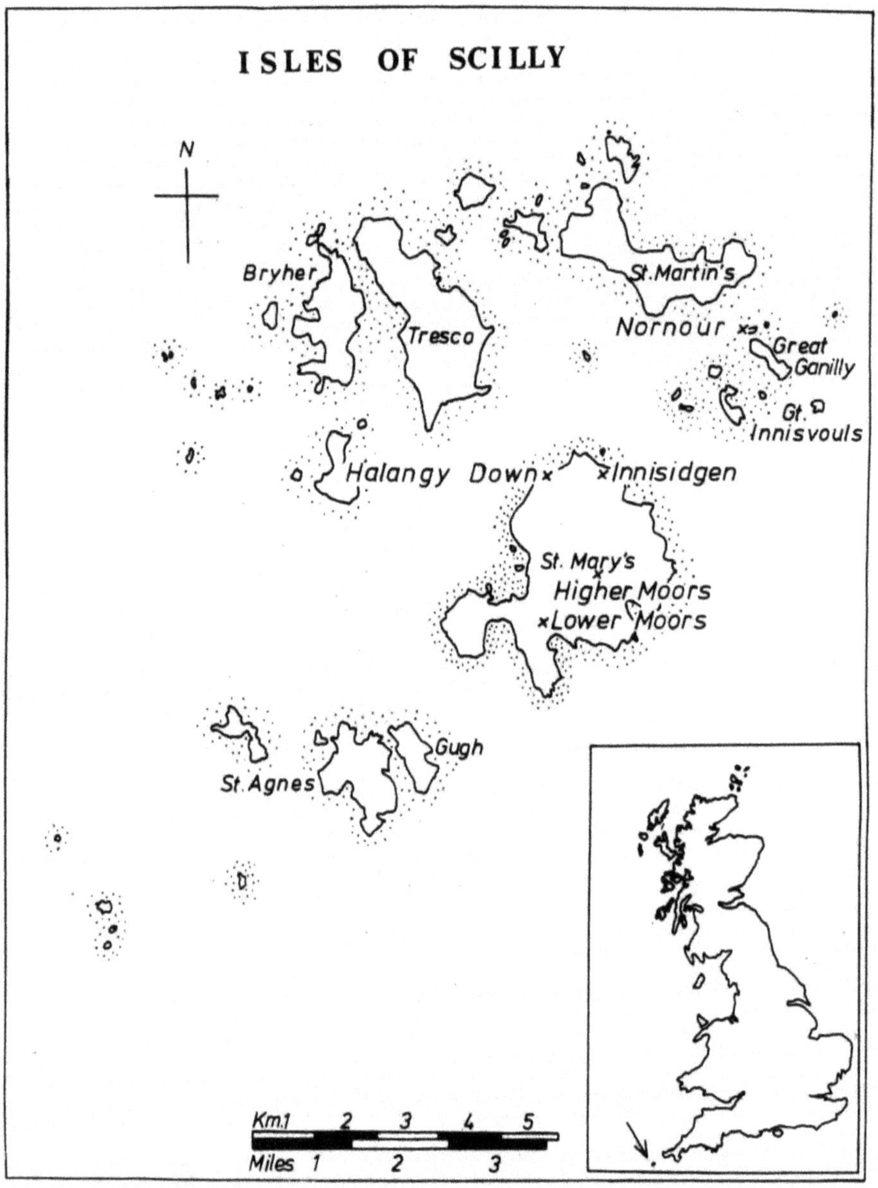

ISLES OF SCILLY
N
Bryher
Tresco
St. Martin's
Nornour
Great
Ganilly
Gt.
Innisvouls
Halangy Down
Innisidgen
St. Mary's
Higher Moors
Lower Moors
Gugh
St. Agnes
Km. 1 2 3 4 5
Miles 1 2 3

Fig. 1

carr with the sedge Carex paniculata to mesotrophic sedge and reed fen and small areas of acid Sphagnum bog. Purely aquatic habitats are present in the numerous drainage channels and in Porth Hellick Pool, itself the result of peat cutting (Lousley, 1971). The least disturbed area of peat was found adjacent to Holy Vale (SV921112) in close proximity to a small area of Ulmus glabra (wych elm) woodland.

A maximum thickness of 76 cm of black, highly humified detritus and monocotyledonous peat was sampled at 4 cm intervals. These organogenic deposits rest sharply on a clean, white, fine to medium grained sand with extremely low organic content. Standard pollen concentration techniques (Faegri and Iversen, 1974; Moore and Webb, 1978) were used in the preparation of 1 ml samples. A pollen sum of 500 grains excluding Filicales and Bryophyte spores was counted. Larger totals were recorded between 4 and 12 cm where local pollen influx from Umbelliferae was high. These results are expressed in pollen diagrams (Figs. 2 and 3) as a percentage of total pollen (TP). The diagrams presented are, however, in abbreviated form showing only the principal types of some 104 pollen and spore taxa recorded. Full pollen diagrams from the site with radiocarbon dates will be presented later, along with comparative material from Lower Moors, St. Mary's. For the purpose of discussion the pollen diagram has been divided into five biostratigraphical assemblage zones which have been designated HM:1-5 from the base to the mire surface. These are characterised as follows:-

HM:1 76-70 cm. Arboreal (53% TP) and shrub pollen (18% TP) dominate this zone with Quercus (oak) (42% AP), Betula (birch) (53% AP), Corylus (hazel) (9% TP) and Salix (willow) being the main elements. Minor records of Pinus (pine), Ulmus (elm) and Alnus (alder) are evident. Herbaceous pollen frequencies are low (31% TP) consisting primarily of Cyperaceae (sedges) (28% TP) and other mire taxa.

HM:2 70-62 cm. This zone is characterised by a reduction in the aboreal components and increased herbaceous percentages. Betula and Quercus decline from previous levels to 17% TP and 3-5% TP respectively. Corylus and Salix remain more or less constant with minor shrub occurrences. Calluna (heather) is incoming and expands to a maximum 6% TP. Herb taxa predominate (62-64% TP) with Gramineae (grasses) (18% TP) and Cyperaceae (34% TP) most important. The presence of pollen of Cerealia and of weeds and ruderals (i.e. plants of disturbed habitats) is of note.

HM:3 62-50 cm. Betula increases sharply (46% TP) subsequently followed by Quercus (10% TP). Corylus and Salix maintain similar levels but with a single maximum of the latter (8% TP). Herbaceous totals show a corresponding decline to 45% TP but with continuation of taxa noted previously.

HM:4 50-4 cm. A sharp reduction in arboreal and shrub pollen totals occurs. Betula and Quercus decline to 5% TP and less than 2% TP respectively. Salix and Corylus similarly decline to less than 2% TP. Conversely, herb totals become dominant (80-90% TP). Maximum pollen frequencies of Gramineae (44% TP) and Cyperaceae (55% TP) are associated with a diverse assemblage of ruderals. Pollen of Umbelliferae

HIGHER MOORS

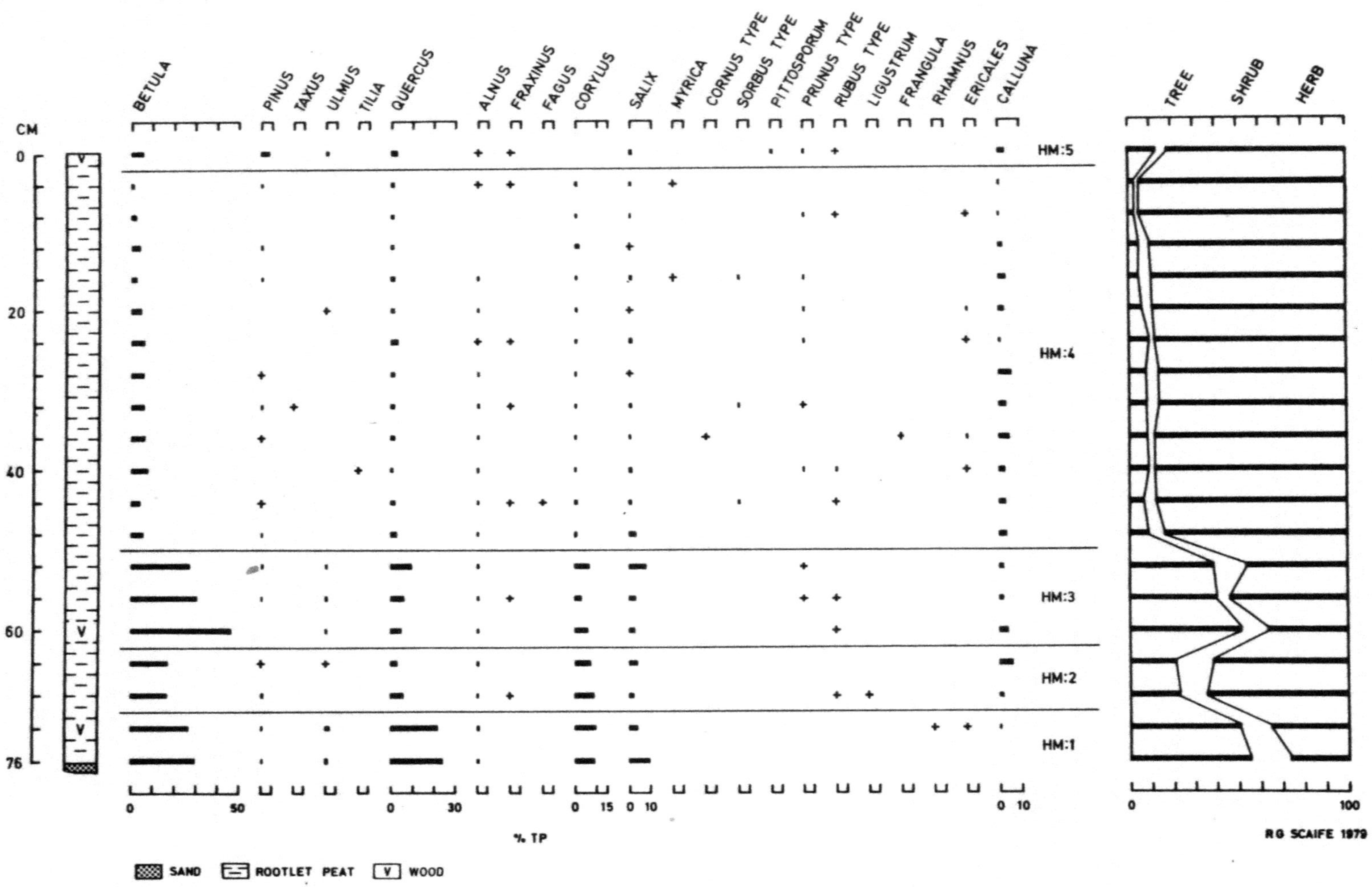

Fig. 2

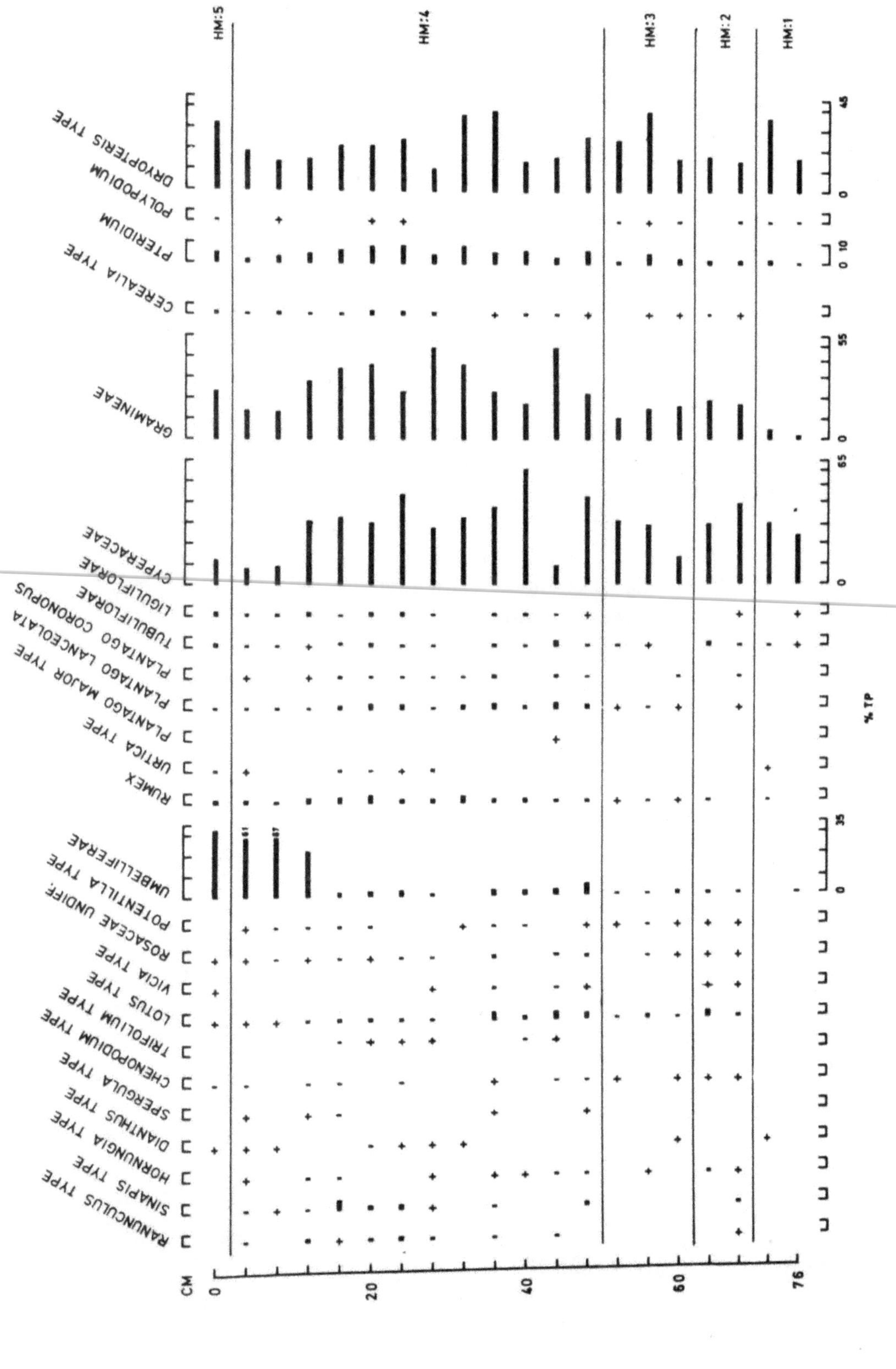

DRYOPTERIS TYPE
POLYPODIUM
PTERIDIUM
CEREALIA TYPE
GRAMINEAE
CYPERACEAE
LIGULIFLORAE
TUBULIFLORAE
PLANTAGO CORONOPUS
PLANTAGO LANCEOLATA
PLANTAGO MAJOR TYPE
URTICA TYPE
RUMEX
UMBELLIFERAE
POTENTILLA TYPE
ROSACEAE UNDIFF.
VICIA TYPE
LOTUS TYPE
TRIFOLIUM TYPE
CHENOPODIUM TYPE
SPERGULA TYPE
DIANTHUS TYPE
HORNUNGIA TYPE
SINAPIS TYPE
RANUNCULUS TYPE
HM:5
HM:4
HM:3
HM:2
HM:1
CM
0
20
40
60
76
% TP

Fig. 3

expands markedly at 12 cm, attaining a maximum of 68% TP after which it remains important. Expansion of ericaceous taxa and Pteridium (bracken) are noted.

HM: 5 0-1 cm. This zone reflects the uppermost peat level, containing slightly higher AP percentages and some exotic pollen. Pinus (8% TP excluding Umbelliferae) and Betula (8% excluding Umbelliferae) expand. Pittosporum and other exotic types are noted. Herbaceous pollen still remains dominant, in the surface level of the mire, dominated by Gramineae (28% TP) and autochthonous Umbelliferae (32% TP).

Discussion

The basal zone (HM: 1) undoubtedly shows the presence of woodland on St. Mary's with the major elements Quercus, , Betula and Corylus dominating the pollen spectra. It is probable that Quercus woodland with Corylus understorey formed the main plant community although the presence of hazel pollen implies a rather open canopy allowing flowering to take place. Betula was also abundant and it may be postulated that this formed regeneration of scrub in areas previously subject to anthropogenic disturbance, scrub growing in exposed coastal localities or, as with Corylus, as an understorey to Quercus. Pinus, Alnus and Ulmus values are low during this phase. Long distance transportation seems the most likely origin for Alnus and Pinus, although Ulmus, having a lesser pollen productivity, may have been localised in Scilly. That little cleared land was present is shown by the relative absence of ruderal pollen taxa. Gramineae frequencies are low while Cerealia type, Plantago (plantain) spp. heathland taxa and other indicator pollen are notably absent. From this evidence, little prehistoric activity, at least adjacent to the site, is postulated during this early forested phase.

At 70 cm in zone HM: 2 strong evidence for anthropogenic activity is seen. A reduction in pollen frequencies of Quercus, Betula and to a lesser extent Ulmus takes place indicating some clearance of the forest. The fact that Corylus remained little changed indicates that either hazel stands were little affected or that opening of the canopy created greater flowering potential, thus maintaining previous levels. New plant communities were created by the forest clearance. Pollen of Calluna and Pteridium marks the initiations of heathland possibly caused by fire and/or soil deterioration. The presence of Cerealia type pollen and associated land use indicators—Cruciferae, Chenopodium type, Artemisia (mugwort) and Anthemis type, all weeds of disturbed ground—show the introduction of arable agriculture. If this cultivation had been of any great extent, higher pollen frequencies of these taxa would be expected. It is probable therefore that cereal cropping was of restricted nature within the pollen catchment. Increased frequencies of Gramineae and of other ruderals typical of disturbed ground and pastoral agriculture occur so that forest clearance for a mixed economy may be postulated.

During this phase of anthropogenic activity, substantial areas of woodland remained in Scilly. With cessation of agricultural activities, a period of woodland regeneration is seen. In zone HM: 3 a marked expansion of Betula to higher values than in HM: 1 and subsequently of Quercus occurs. A reduction in *Corylus* may imply that hazel was being shaded out by the re-establishment of *Quercus* woodland, or failed to flower under oak canopy. Alternatively the effect may be due to differential filtration of pollen into the catchment (Tauber, 1967). It is apparent, however, from the Cerealia type pollen and herbaceous elements still present, albeit in low frequencies, that much open land remained.

At 50 cm, a very marked decline in arboreal pollen frequencies occurs after which woodland taxa play subordinate roles to the herb flora. This horizon (HM:3/HM:4 junction) is problematical in that it may represent one of three situations. These are:-

a) Cutting for peat fuel has occurred in most of the two mire areas. Truncation of the profile may thus have occurred with removal of peats representing a substantial period of time.

b) A hiatus may have occurred due to the cessation of peat formation caused by climatic and/or hydrological agencies.

c) Extensive deforestation with *Quercus*, *Betula*, *Corylus* and *Salix* being removed from the landscape.

In view of the highly humified nature of the peat and the absence of any contrast between markedly older and younger peats, cutting is not thought a likely cause. Similarly, no stratigraphical change was noticed at this horizon which would indicate any break in organic deposition. The humified nature of the peat may, however, mask any such change and might be a possible cause so that woodland may have been cleared during the hiatus. The final hypothesis seems mostly likely in view of the stratigraphy and the corresponding increase in herb taxa. It is certain, however, that zone HM:4 represents a great period of time during which arboreal vegetation was notably absent. Expansion of ruderals, including *Plantago lanceolata* (ribwort plantain), *P. coronopus* (buckshorn plantain) and Gramineae, shows the openness and pastoral nature of the landscape. At 28-16 cm minor increases of Cerealia type and *Sinapis* type show that a phase of arable activity occurred. This possibly represents cultivation of areas closer to the mire rather than any widespread extension of cultivation. A change in the status of the mire, possibly brought about by drainage, resulted in the rapid expansion of Umbelliferae. At 12 cm the marked increase in Umbelliferae (68% TP) pollen corresponds to the autochthonous growth of *Oenanthe crocata* (hemlock water dropwort) still present on the site.

The uppermost level of the mire (HM:5) displays the vegetation changes that have taken place during the last 150 years. A minor increase in *Pinus* reflects the planting of *P. radiata* in a number of localities in the islands (Lousley, 1971). Similarly low frequencies of exotic taxa were found. These were in lower numbers than expected and with the exception of *Pittosporum* (2 grains) no attempt was made to identify those present. Historical changes and the introduction of exotic types into Scilly have been described by Lousley, providing more accurate information than can be given by pollen analysis.

A POLLEN ANALYSIS FROM NORNOUR (JRAG)

The Archaeological Background

A settlement occupied in prehistoric and Roman times has been excavated on Nornour (grid ref. SV944148), a small island in the Eastern group of the Isles of Scilly (Dudley, 1968; Butcher, 1970, 1971, 1972, 1974, final report nearing completion). The work was sponsored by the Department of the Environment. In 1972 the slope of the island behind the settlement was investigated in a series of trenches. The samples discussed in this note were taken from the highest trench, LI, which was about 6 m south of the crest of the island and 30 m north of House 6.

All these trenches showed the same general stratification; a deposit of blown sand lying over a thin soil covering granite rocks. There were no artifacts or other signs of human activity. The sand covers the whole of the south side of Nornour, including the ancient settlement, and the soil seems to correspond to the ground surface on which the buildings stand. However, since it would have been open during the occupation of the site this gives a very wide chronological range. The earliest radiocarbon dates are before 1,000 b.c. and occupation continued until the fourth century A.D. There is no evidence as to how long the surface remained open after the settlement was finally abandoned but it seems reasonable to suggest that the sand began to accumulate soon afterwards.

The Soils

On Nornour soil parent material appeared to consist of locally derived head deposits of gravelly loam with variable stone content, overlying granite rock, simllar to that found on the granites in Cornwall (Clayden, 1964). The soil buried beneath the regosol derived from blown sand on the south side of Nornour was a ranker with a shallow A/C profile approximately 30 cm in depth, and it appeared to be similar to the soil associated with the ancient settlement. The site was on a slope of 10° and freely drained. The vegetation was predominantly *Pteridium aquilinum* (bracken) and earthworms were absent from the profile. The section sampled for pollen analysis is described by Dr. Helen Keeley as follows:-

0-22 cm (the F horizon): A dark brown (10YR3/2) fibrous organic layer containing bulbs of *Endymion non-scriptus* (bluebell), bracken roots and undecomposed and decomposing plant debris. Sand grains were present.

22-32 cm: Grey (10YR6/2) structureless sand containing roots and small particles of amorphous organic matter, probably washed down from the F horizon.

32-42 cm: Structureless sand, stained dark grey to black by eluviated organic matter (10YR3/1) containing bracken roots. There may have been some consolidation of the blown sand at this stage of development.

42-62 cm: A light coloured (10YR6/3) wind-blown sand; structureless and containing bracken roots.

62-86 cm: The buried A horizon of the original soil on which the sand was deposited. This layer consisted of a black (7.5YR2.5/0) silty sand, with weak crumb structure, containing amorphous organic matter and few roots and was more compact than the layers above.

86-91 cm: (bC horizon) was dark brown (7.5YR3/2) rotted (weathering) granite, containing granite fragments, with coarse loamy aand texture and weak crumb structure. Roots were absent and the organic matter content was much less than in the layer above.

91 cm -: Granite boulders occurred below 91 cm.

The Pollen

The upper part of the profile where the soil was light and sandy was sampled every 5 cms, but below 62 cm where it was darker and more organic the sampling interval was reduced to 2.5 cm. The samples consisted mainly of sand, and the organic matter was extracted by standard treatment. The black humus-stained sand from below about 62 cm was pollen rich but the soluble humic matter had to be removed first by means of repeated alkali washes and centrifugation. Treatment with hydrofluoric acid was carried out in all cases, but acetolysis was found to be unncessary.

There was abundant pollen in most of the samples, but counting was made difficult by the fact that it was usually very shrunken and crumpled and the abundance of thick-walled pollen grains in some samples suggests differential preservation. Counts of 200-500 pollen grains were made and although absolute counting was not done, an estimate of pollen abundance was made, recorded on the right hand columns of the pollen diagrams. The pollen diagram has been drawn up on the basis of percentages of total pollen (tree pollen + herbaceous pollen), and it has been somewhat simplified by the exclusion of some sporadic pollen records.

The pollen diagram (Fig. 4) can be divided into three sections on the basis of pollen frequencies as follows:-

10-30 cm: Characterised by high Gramineae, Ericales, *Armeria* (thrift), more Compositae—Tubuliflorae than Liguliflorae. Moderate pollen content.

35-60 cm: High Compositae—Liguliflorae, Caryophyllaceae, Umbelliferae, Cyperaceae. Low Gramineae, Ericales, *Armeria*, tree pollen. Low pollen content.

62.5-87.5 cm: High *Plantago*, Gramineae, moderate Ericales, Cerealia, tree pollen, *Armeria* present. Low Compositae and both types nearly equal, practically no Umbelliferae or Cyperaceae. Very high pollen content.

Interpretation

The top section of the pollen diagram (10-30 cm) is dominated by herb pollen, mainly that of grasses, plantains and heather and the topmost sample contains significant numbers of bracken spores. This part of the diagram represents to some extent the pollen rain from the present day vegetation growing in the vicinity, although pollen rain is necessarily an incomplete record owing to the vagaries of pollen production and dispersal from plants, its

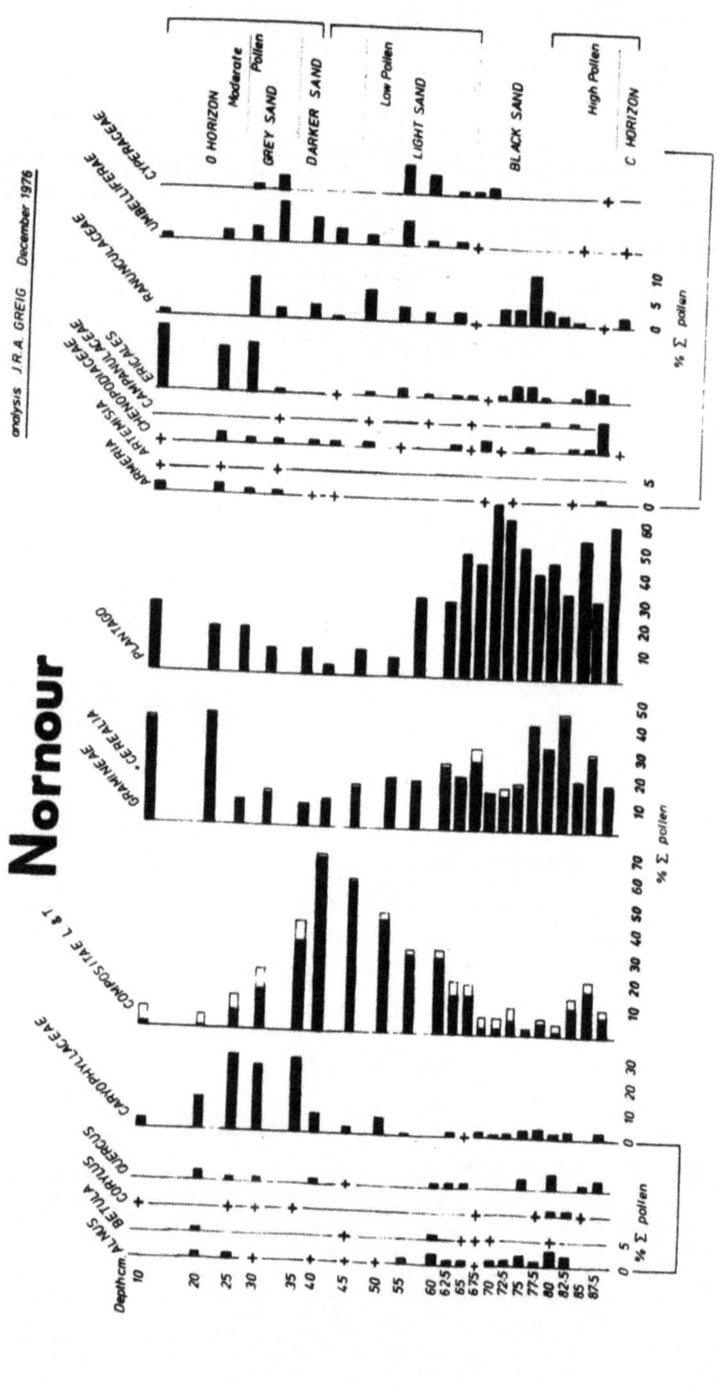

Nornour
analysis J.R.A. GREIG December 1976
Depth cm
ALNUS
BETULA
CORYLUS
QUERCUS
CARYOPHYLLACEAE
COMPOSITAE L & T
GRAMINEAE
CEREALIA
PLANTAGO
ARMERIA
ARTEMISIA
CHENOPODIACEAE
CAMPANULACEAE
ERICALES
RANUNCULACEAE
UMBELLIFERAE
CYPERACEAE
% Σ pollen
0 HORIZON
Moderate Pollen
GREY SAND
DARKER SAND
Low Pollen
LIGHT SAND
BLACK SAND
High Pollen
C HORIZON

Fig. 4

preservation and its final identification. Sometimes vegetation and record are linked as in the case here where there are bracken rhizomes in the ground and also a corresponding record of spores but in other cases this is not so. There were bluebell bulbs in the soil as well but their pollen is fragile and shed in small quantities and was not found in any preparation, so this study can shed little direct light on the history of bluebells here.

The _Flora of the Isles of Scilly_ (Lousley, 1971), is helpful in linking the pollen record with the various plant communities which make up the flora of Nornour, mainly variants on the Down and Dune types of vegetation, and most of the herbaceous pollen can be accounted for in this way. The Caryophyllaceae pollen could not be identified more exactly, but on the basis of recent records _Silene maritima_ (sea campion) or one of the species of _Cerastium_ (mouse-ear) or of _Stellaria_ (chickweed) are candidates. The Compositae are also a large group, and the section Tubuliflorae pollen which dominates this upper part could be from _Achillea millefolium_ (yarrow) or _Tripleurospermum maritimum_ (sea mayweed). A species of _Leontodon_ (hawkbit) could be responsible for the Compositae sect. Liguliflorae pollen record. Gramineae pollen is the most abundant type in this part of the diagram, difficult to attribute to any one species, but the _Plantago_ pollen could mostly be attributed to _P. Coronopus,_ a major component of the maritime grassy sward. _Armeria maritima_ is another specific pollen record which implies the existence of seaside vegetation, and the Ericales record probably represents the heather (_Calluna vulgaris_) which is a major part of the natural plant cover and more widespread than _Erica cinerea_ (fine-leaved heath). The Ranunculaceae pollen would appear to have come from _Ranunculus ficaria_ (lesser celandine), the only member of this family known on the Eastern Isles, and _Carex arenaria_ (sand sedge) is likewise the most abundant and widespread member of the Cyperaceae here.

The herbaceous pollen record appears to correspond to what might be expected from the present day flora, and plants which release quantities of identifiable pollen seem to be represented, although there are others which are abundant on the ground like bluebells and perhaps brambles which do not feature here. The tree and shrub pollen record is much smaller than that from the herbs but _Alnus_, _Betula_, _Carpinus_ (hornbeam), _Corylus_ and _Quercus_ are represented, although the topmost sample has only a single grain of hazel pollen. It is not easy to see the parallels between this part of the pollen diagram and the regional flora because Scilly is today virtually treeless. Alder and oak grow in a few places, but birch, hazel and hornbeam do not (Lousley, 1971) and Nornour itself is treeless. The _Flora_ does point out that one of the Eastern Isles close to Nornour, Great Ganinick, has many plants growing there which would normally be associated with woodland, and it is the only place apart from Tresco where oak has been recorded, so these may be signs of past woodland.

There are several possibilities which might explain the occurrence of these three apparently exotic tree pollen types in this part of the profile. The first is that the sand just below the present ground surface may have been deposited some time ago with, perhaps, little build-up since and the signs of woodland represent something that existed previously. An alternative is that the soil profile may have been truncated at some time by wind erosion, thus exposing older deposits on which the present ground surface formed. Another factor is the deposition of sand eroded from elsewhere which could result in the mixing of some older deposits with younger ones, for blown sand seems to be an important factor in Scillonian soil formation.

The next pollen assemblage (35-60 cm) would appear to represent a somewhat different vegetational type from that above it but the pollen was generally poorly preserved. Some of the changes, like the superabundance of Compositae (L) pollen, could be due to differential preservation and thinner walled pollen grains may have become corroded beyond recognition. Without the additional information from absolute counts it is difficult to be sure whether the abundance of Compositae (L) pollen is mainly due to preservation factors. Even so, there are discernible changes such as the smaller amounts of pollen from Armeria and Ericales which are robust pollen grains too and unlikely to be affected by preservation factors. This pollen assemblage would appear to correspond with the phase of consolidation of blown sand mentioned in the soil report, and the low pollen values could well correspond with this observation with an unstable dune surface supporting less of the heather and thrift.

The next pollen assemblage (62.5-87.5 cm) corresponds with the buried soil surface and has the highest pollen values. The most abundant pollen type is Plantago which with the large amount of Gramineae pollen would suggest a kind of maritime sward vegetation, maybe maintained by grazing pressure which would favour plants with basal rosettes like grasses and plantains at the expense of others not so adapted to tolerate grazing. There are occasional records of pollen considered to be that of Cerealia. It is rather a sporadic record, but the largest amounts occur in this lowest part of the pollen diagram and provide evidence of arable farming as well as the possible pasture land suggested by the plantain and grass pollen. The blackness of this soil may be due to the use of seaweed, spread on the fields to counteract the 'hungry' quality of sand soil, but of course no direct traces of this could survive.

The tree pollen record is at its most consistent, with almost continuous occurrences of Alnus and Quercus, more scattered ones of Betula and Corylus and perhaps traces of Ulmus which amounts to evidence that there was at least some woodland within range of pollen dispersal at this stage.

This part of the pollen diagram may correspond with the occupation of the Nornour site, on stratigraphic grounds, although the dating for the latter is not very precise, 1,000 B.C. - 400 A.D. It may be added that in 1966 Dimbleby (personal communication) examined the buried soil beneath the buildings for pollen but none was found.

Discussion

An interesting aspect of these results is the comparison between the plants appearing in this pollen diagram from Nornour and other Scillonian pollen work, those known from fossil macroscopic remains (mainly charcoal) and the present day flora described in the Flora of the Isles of Scilly. The known fossil records of ash, hornbeam and perhaps elm are supplemented by the evidence from Nornour. These trees are not nowadays native on Scilly, but their past presence is shown by a charcoal record of ash (Flora, p. 212), but it could be argued that charcoal records might represent wood which had been imported while pollen would almost certainly demonstrate the presence of plants growing

Scilly. The hornbeam and elm pollen records come from Dimbleby's results from Innisdgen. Other interesting pollen records include those of Artemisia, Dipsacaceae and Filipendula which represent plants which are not known to grow on the Isles of Scilly now.

Some plants may have escaped detection by pollen analysis yet were probably abundant at times, particularly Ulex (gorse). This family, the Leguminosae, generally produces pollen that is thin and frail and also tends to be hard to identify to any group of genera within the family even when fresh. It is not surprising that the small amounts of Leguminosae pollen found in five samples could not be positively attributed to Ulex. There have been several reports on charcoal from Nornour. The most recent is that of Mrs. C. Keepax (Ancient Monuments Laboratory Report No. 2029) in which she identified most of the charcoal as Leguminosae. It was not, however, possible to tell with certainty from the charcoal whether this was Ulex or not, even though gorse is the most likely possibility as it is a traditional source of fuel and grows abundantly on Scilly.

The lower part of the pollen diagram would appear to correspond to a time when the site was occupied. The signs of arable farming of cereal crops should not be taken to mean that these were necessarily the only crops grown for few other crops produce abundant and identifiable pollen but there do not appear to by any records of carbonised plant remains such as grain to confirm or even amplify this picture. The evidence of pastoral farming is corroborated to some extent by the animal remains which also demonstrate that the sea, with its birds, animals and fish, would appear to have provided useful food as well (Turk, 1971).

HALANGY DOWN, ST. MARY'S (GWD)

The Iron Age settlement of Halangy Down was excavated from 1964 onwards by Ashbee (1964-1970). No clearly defined palaeosols could be found beneath any of the structures on this site from which soil pollen profiles might have been studied. Samples immediately overlying some of the structures did prove to be pollen-rich but of course they could not relate to contemporary land use. The only point worth making about them is that they were virtually devoid of the pollen of woody species; Gramineae and Pteridium predominated, with varying amounts of Calluna and a strong suite of ruderals.

In the hope of obtaining a soil pollen profile that might cover the period of occupation of the Halangy site, a series of samples was taken in the marine cliff lying some 150 yards downslope from the site. The cliff section showed about four feet of unconsolidated deposits, possibly hillwash, overlying a ten foot depth of unmodified 'head'. (At this date, 1966, records were made in imperial, not metric units).

Contiguous samples at one inch intervals were taken through the unconsolidated deposits at the top, and also in a loamy layer, possibly a weathered soil, immediately beneath the 'head'. This latter proved to be devoid of pollen and need not be mentioned further.

The pollen diagram obtained from the upper layers (excluding a few 'casuals') based on total pollen plus fern spores, is shown in Fig, 5 and alongside it the stratification of the deposit.

Interpretation

The most striking feature of this pollen profile is the virtual absence of tree pollen. This means that there is no means of dating the different layers by tree species. The only date indicator is a piece of 14th century pottery found at ten inch depth, in the surface layer of loose grey sand. This probably implies, though it does not prove, that the layer below this in the profile is of earlier date.

The pollen diagram falls into five distinct layers, indicated by horizontal lines across the diagram. The topmost layer, only two inches deep, is characterised by a high proportion of grass pollen, and particularly of the group described in the diagram as Varia-Miscellaneous. These are pollen types which could not be identified (though well preserved) and probably represent aliens associated with the present use of the land for commercial flower growing. The pollen of *Pittosporum*, an introduced evergreen shrub used today as a windbreak, was found only in these two samples.

From two to nine inches, the profile is dominated by grass, Liguliflorae and ruderal pollen, together with abundant bracken spores. The fact that the main curves show distinct trends implies that pollen is being carried down into this layer, so that despite the loose nature of this material it had not been added to in recent times but had developed a stable surface from which a sequence of pollen spectra has developed.

This is in marked contrast to the underlying layer, in which the pollen is evenly distributed both in frequency and percentage. It would be reasonable to regard this as a layer of accumulating top soil, probably hillwash from the fields above. The pollen suite is poor in species and is largely dominated by Liguliflorae and bracken. As both of these pollen types are relatively resistant to decay, we may be seeing here the relic of a richer assemblage. At 23 inches, a sudden change occurs, marked by higher percentages of grass pollen and much less bracken. The most conspicuous feature, however, is the sudden increase in the pollen of Cruciferae, of a type which could not be identified for certain, but had a close resemblance to *Cakile maritima* (sea rocket). The grains are smaller than modern specimens, but this is commonly so with pollen preserved in mineral sediments. The Cruciferae, of course, include crop plants such as cabbage or charlock, so the significance of this pollen type must remain in doubt. However, I know of no parallel for high concentrations of the pollen of this family, which is entomophilous, and its presence must mean the presence of quantities of flowering material. Possible explanations are that either a flowering crop was being grown as a green manure of that local coastal vegetation was being brought on to the fields for the same purpose.

Whatever the explanation, the distribution of this pollen type in this layer suggests that at 23 inches there is an old surface subsequently buried under hillwash. This layer itself, however, is clearly not a primary soil, for at 31 inches another apparent surface appears, also showing high values of Cruciferae pollen but characterised by a new pollen, type, so far not identified. Re-examination of this pollen type suggests that it is also a crucifer, possibly *Cochlearia anglica* (scurvy-grass), another coastal plant. However, it clearly distinguishes the lowest of the five layers. This has all the indications of a pollen profile developed in a soil *in situ*, which is confirmed by the description

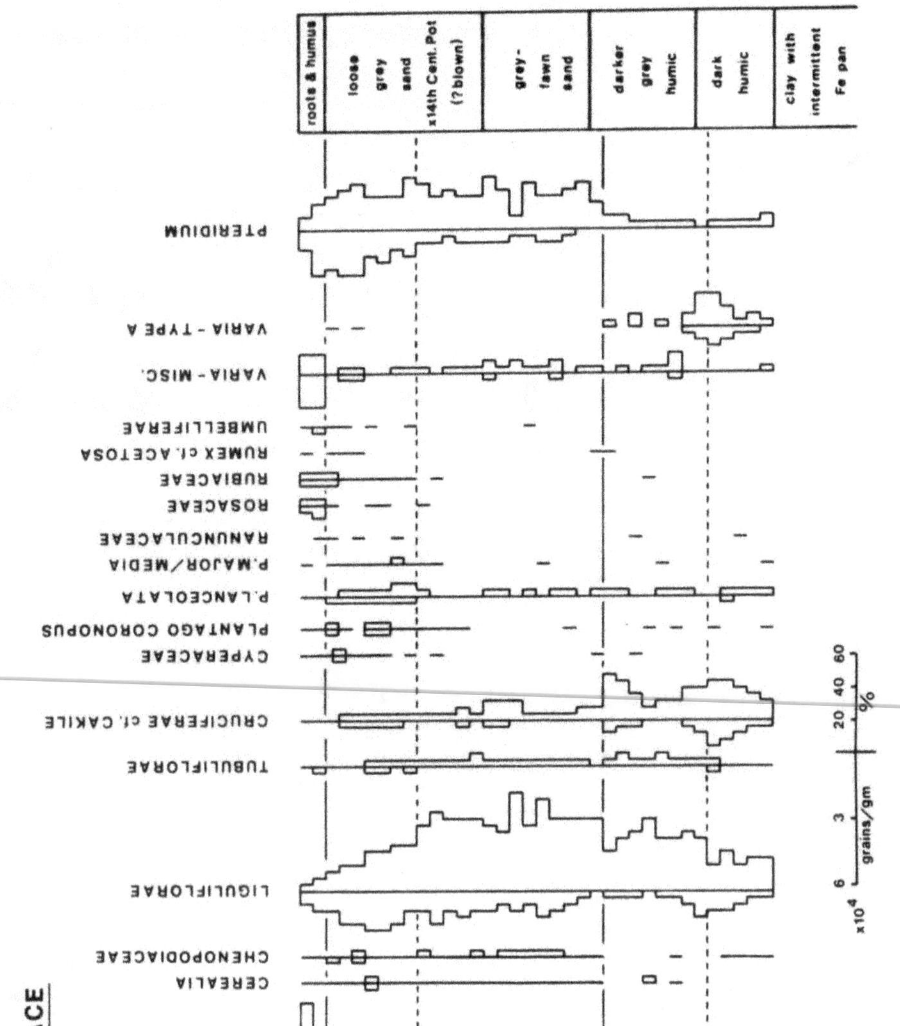

HALANGY DOWN-CLIFF FACE
BETULA
PINUS
QUERCUS
CORYLUS
CALLUNA
GRAMINEAE
CEREALIA
CHENOPODIACEAE
LIGULIFLORAE
TUBULIFLORAE
CRUCIFERAE cf. CAKILE
CYPERACEAE
PLANTAGO CORONOPUS
P. LANCEOLATA
P. MAJOR/MEDIA
RANUNCULACEAE
ROSACEAE
RUBIACEAE
RUMEX cf. ACETOSA
UMBELLIFERAE
VARIA - MISC.
VARIA - TYPE A
PTERIDIUM
roots & humus
loose grey sand
x14th Cent. Pot (? blown)
grey-fawn sand
darker grey humic
dark humic
clay with intermittent Fe pan
0
5
10
15
20
25
30
36 ins.
x10^4 6
3
grains/gm
20
40
60
%

Fig. 5

of the lowest two layers shown at the right-hand side of Fig. 5. It seems unlikely that this soil was covered by deposits from close by, for the samples above it give mirror images of the pollen curves, especially the Cuciferae and the Varia-Type A.

It is apparent that though this level of 29 to 36 inches is the lowest polleniferous horizon, there is no greater representation of tree pollen in it than in any of the layers above. In this sense, it contrasts with the Nornour profile which is both somewhat earlier and on a small island. However, the bog evidence from Higher Moors and the palaeosol evidence from Innisidgen, neither of which is more than a mile from Halangy Down, clearly establish the previous presence of woodland in this area of St. Mary's. It therefore has to be assumed that Halangy Down itself was once wooded. Though it is not possible with certainty to equate the lowest buried levels of the cliff face pollen profile with the Iron Age occupation above, in the absence of any other archaeological settlements in the vicinity this would be a likely explanation. If so, not only were the environs treeless at this time but apparently all traces of a primary soil have disappeared, presumably by hillwash or wind erosion. It will be remembered that the palaeosol at Innisidgen itself held evidence of truncation of the primary soil profile and the superimposition of blown sand associated with agricultural land use.

CONCLUSION

It is impossible to base a detailed interpretation of the vegetational history of the Isles of Scilly on the few investigations so far made, but even from what has been done one or two general conclusions can be drawn. There is now clear evidence that the large island of St. Mary's had a cover of deciduous woodland sufficiently continuous in canopy to exclude light-demanding herbs. The Nornour pollen analysis indicates greater influence of trees in the lower layers, but it remains an open question whether this was due to tree growth on this small island itself or whether it was a reflection of the conditions on adjacent larger islands.

On the Cornish mainland deforestation started in earnest in the Bronze Age (Dimbleby, 1963 and 1971) and if this were the case on the Isles of Scilly, it would agree with the signs of a modified landscape during and after the occupation of the site on Nornour. The pollen profile from Higher Moors provides clear evidence of an extensive clearance which could correlate with such occupation, and the evidence of clearance found in buried soil at Innisidgen could also be associated with this phase. In the absence of time markers in any of the pollen diagrams, certainty about dates will have to await the radiocarbon dates from the Higher Moors peat profile.

If the evidence of agriculture at Halangy Down can be attributed to the adjacent Iron Age settlement, that area of St. Mary's was completely treeless by that time. There is evidence both from this site and from Nornour that prehistoric agriculture was leading to soil movement both by hillwash and by wind erosion. If there is an exposure effect on such off-shore islands it may express itself not so much in the nature of the primary vegetation as in site deterioration following the disruption of the original woodland cover by prehistoric agriculture and through its continuing use.

ACKNOWLEDGEMENTS

The authors are grateful to Miss S. A. Butcher for providing the archaeological information about the site on Nornour and to Dr. Helen Keeley for her description of the soils there.

BIBLIOGRAPHY

Ashbee, P., 1965-1970. Excavations at Halangy Down, St. Mary's, Isles of Scilly. Cornish Archaeology 4, 1965, 36-40; 5, 1966, 20-27; 7, 1968, 24-32; 9, 1970, 69-76.

Butcher, S. A., 1970. Excavations at Nornour, Isles of Scilly, 1969-70. Cornish Archaeology 9: 77-91.

Butcher, S. A., 1971. Excavations at Nornour, 1971. Cornish Archaeology, 10: 94.

Butcher, S. A., 1972. Excavations at Nornour, 1972. Cornish Archaeology, 11: 58-9.

Butcher, S. A., 1974. Nornour. Isles of Scilly Museum Publication No. 7.

Clayden, B., 1964. In: Hosking, K. F. G. and Shrimpton, G. J. (eds.), Present views on some aspects of the geology of Cornwall and Devon. Royal Geological Society of Cornwall, Penzance, 311-330, Soils of Cornwall.

Coombe, D. E. and Frost, L. C., 1956. The heaths of the Cornish Serpentine. J. Ecol., 44: 226-256.

Dimbleby, G. W., 1963. Pollen analyses from two Cornish barrows. Journal of the Royal Institution of Cornwall n. s. IV, part 2: 364-375.

Dimbleby, G. W., 1971. In: Miles, H. & T., Excavations on Longstone Downs, St. Stephen in Brannel and St. Mewan. Cornish Archaeology, 10: 26-7.

Dimbleby, G. W., 1977. A buried soil at Innisidgen, St. Mary's, Isles of Scilly. Cornish Studies, 4/5: 5-10.

Dudley, D., 1968. Excavations on Nornour in the Isles of Scilly, 1962-66. Archaeological Journal, CXXIV: 1-64.

Faegri, K. and Iversen, J., 1974. Textbook of Pollen Analysis. 3rd edition by Faegri, K., Blackwell, Oxford.

Gimingham, C. H., 1972. Ecology of Heathlands, Chapman and Hall, London.

Keepax, C., 1976. Ancient Monuments Laboratory Report No. 2029.

Lousley, J. E., 1971. Flora of the Isles of Scilly. David and Charles, Newton Abbot.

Moore, P. D. and Webb, J. A., 1978. An Illustrated Guide to Pollen Analysis. Hodder and Stoughton, London.

Tauber, H., 1967. Differential pollen dispersion and filtration. Proc. Cong. Int. Ass. Quatern. Res., 7: 131-41.

Turk, F. A., 1971. Notes on Cornish mammals in prehistoric and historical times: 4. Cornish Archáeology, 10: 79-91.

DIATOM ANALYSIS FROM COASTAL SITES IN THE NETHERLANDS

By M. J. Jansma

Introduction

In order to understand the analysis of diatoms from ancient settlements in coastal regions better, it is perhaps pertinent to explain first briefly what diatoms are, the environment in which they live, in which type of sediment they may be found in a fossil state, how slides are made from both soil samples and pottery and how the count method has developed.

Diatoms are single-celled plants, either round or long in shape. Their size varies from a few microns to c. 1 mm (= 1000 microns). Their primary requirement is sunlight in order to assimilate carbon dioxide, in the same way as do other plants, and secondly they require water. The water must contain silica in solution which the diatoms extract to form their shells (or thecae). These shells fit into one another like a box and its lid. Both the shells together form one cell. Ecologically, diatoms may be subdivided into three major groups, depending on the salinity of the water which they inhabit, namely, marine, brackish and fresh water diatoms. Temperature, current, depth and acidity of the water are also of importance.

In the Netherlands, fossil diatoms are best preserved in the Holocene coastal deposits, i.e. in clays, which may or may not be mixed with sand or humus, though they are also to be found in more peaty deposits. Away from the coast diatoms occur in riverine and lacustrine sediments. Finally, diatoms may occur in the clays used in pottery manufacture, on condition that the firing temperature never exceded 800°C. The presence of diatoms in pottery was discovered quite accidentally by A. van der Werff.

Turning now to the preparation of microscope slides from both sediment and pottery samples which are thought to contain diatoms, the sample is first boiled up with concentrated hydrogen peroxide which oxidizes away all organic matter with the exception of the diatoms. If the sample is calcareous a couple of drops of concentrated hydrochloric acid are sufficient to dissolve the remnants of calcium carbonate.

The sand in the sample is allowed to settle, after which the liquid, containing clay and diatoms, is poured off. The mixture is centrifuged a couple of times, after which the liquid above the sediment is poured off, removing the clay fraction smaller than 2 microns, though not the diatoms. The residue —consisting of diatoms and the remnants of the clay fraction 2-6 microns— may now be transferred to a slide, enclosed in a synthetic resin with pronounced refractive tendencies which enhance the structure of the diatom shells when viewed under a microscope.

Before making slides from pottery sherds, these should be cleaned with the utmost care since all foreign clay particles must be removed. Poorly fired sherds may then be crumbled by hand, while better fired sherds must be splintered mechanically. The sherds are then treated in the same fashion as the sediment samples.

Concerning the count method, the number of specimens per species is counted for each sample, preferably to at least 400 individuals, just as in pollen analysis.

In addition, it has proved to be of great importance to record the percentage of diatoms per species which are broken and which are whole, since in the study of Feddersen Wierde by Körber-Grohne, it appeared that transport of the diatoms over longer distances tends to inflict damage to the shells, although the total effect is to some extent dependent on the shape and size of the shells.

By registering the proportion of broken to complete individuals it is in many cases possible to establish whether a particular species lived locally, or whether it was transported from elsewhere: these are referred to as autochthonous and allochthonous species respectively. The distinction is especially important because diatom shells can be extremely resistant and may survive several removals in fairly good condition. Diatom shells can also break as a result of drying out.

By collecting and identifying the diatoms at present occurring in the Netherlands van der Werff has considerably enlarged our knowledge of recent species. His work has resulted in the compilation of an identification list of diatoms in the Netherlands (van der Werff & Huls, 1957-1974).

II. Purpose and importance of diatom analysis to settlement archaeology

a. Diatom analysis of sediment samples

Since—as we have already seen—diatoms live in water, the only sediments near to settlements which contain diatoms are those deposited in or by water. The settlement might, for example, be situated on the coast, near a river, creek, lake or stream. Diatoms in soil samples from a settlement may therefore indicate the presence of ditches, runnels or moats in or around the settlement. If the individual diatoms in such sediments are complete, then the clay in which they are found must have been deposited *in situ*, in saline, brackish or fresh water conditions. If, on the other hand, the species are broken, secondary displacement is implied or alternatively, the sediment may have been exposed during dry periods. In reality, there is usually a mixture of species with complete shells with species with damaged shells. Thus the diatom content of a particular sediment sample which consists of unbroken individuals of fresh water species and damaged individuals of marine species might indicate that the environment around the settlement was originally fresh, but that later, during a transgression, scales of marine species were washed in with the sea water. For the inhabitants, the consequences might have been total abandonment of the settlement, or the compulsory raising of the inhabited area in order to remain. A transgression may lead to a situation whereby the lower sediment contains fresh water diatoms while the upper deposit contains marine species. The transgression may in turn be followed by a regression, when the marine species are seen to be replaced by fresh water species etc.

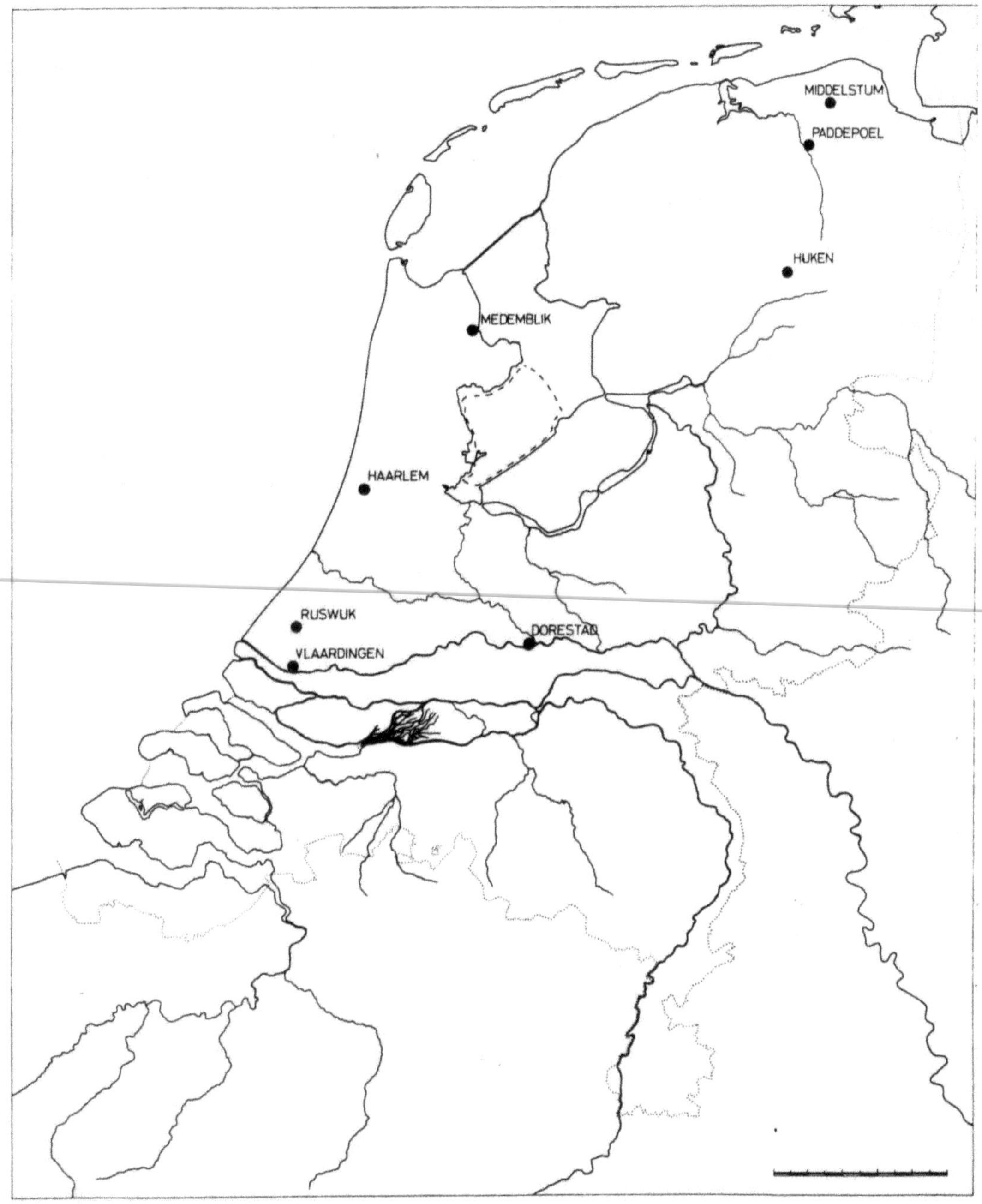

Fig. 1 Map of the Netherlands, indicating coastel sites. The scale in the lower right corner is 50 km.

The investigation at Paddepoel, Groningen (fig. 1) is a suitable example of diatom analysis filling in the picture of the environment in and around a settlement (Groenman-van Waateringe & Jansma, 1968).

A quantitative diatom analysis of two standard profiles (Paddepoel III-330 and Paddepoel IV-345), has been carried out, and the fluctuations in the percentages of the individual species have been diagrammed in the same way as is customary in pollen analysis (fig. 2). The following results were obtained:

1. Prior to the earliest phase of occupation, a gradual decrease in the intensity of flooding, becoming practically zero in the phase of occupation. The diatoms are allochthonous (on the basis of the high percentage of broken specimens), except for a number of brackish-fresh and fresh-brackish species which could have lived in wet places on the high mud-flat.

2. Between the earliest phase of occupation, encountered at Paddepoel I, II and III, and the latest occupation phase, found at Paddepoel IV, an increase in flooding, as may be concluded from the percentage of allochthonous marine plankton which is higher than it was toward the end of the preceding sedimentation phase.

3. At the time of the second occupation, a decline inflooding activity, which strongly increased thereafter. A quantitative analysis of a ditch section from Period I at Paddepoel III (Paddepoel III-314) gave a picture quite different in character from the analysis of the standard diagram.previously shown. In the latter, the diatoms are for the most part allochthonous, transported from elsewhere and deposited with the clay sedimentation and in the occupation phases, when there was no clay deposition, a limited range of a brackish-fresh to fresh-brackish diatom flora occurred in wet places on the mud-flat. In the ditch however, a predominantly autochthonous diatom flora occurred, with a brackish to fresh-brackish character at the bottom and brackish-fresh to fresh higher up. The small quantities of marine to brackish-marine species are allochthonous (fig. 3).

A second example is drawn from the Roman period settlement near Rijswijk in South Holland (fig. 1). In this settlement numerous ditches were found (Bloemers, 1978). When the fill of some of these ditches was examined for diatoms it became clear that two types of ditches were distinguishable (Jansma, 1978);

1. enclosure ditches directly around the settlement which were filled with all manner of household rubbish,

2. boundary ditches, enclosing plots which were proved by pollen analysis to be arable fields (Groenman-van Waateringe, 1978).

The ditches of this last category were in open connection with the enclosure ditches, though about 10 cm lower. The diatom content of both ditch categories is broadly similar, but in the enclosure ditches virtually all the diatoms are damaged, while in the field ditches only the marine and marine-brackish species occur, with a high percentage of damaged specimens. This

Fig. 2 Paddepoel III-330 (standard diagram).

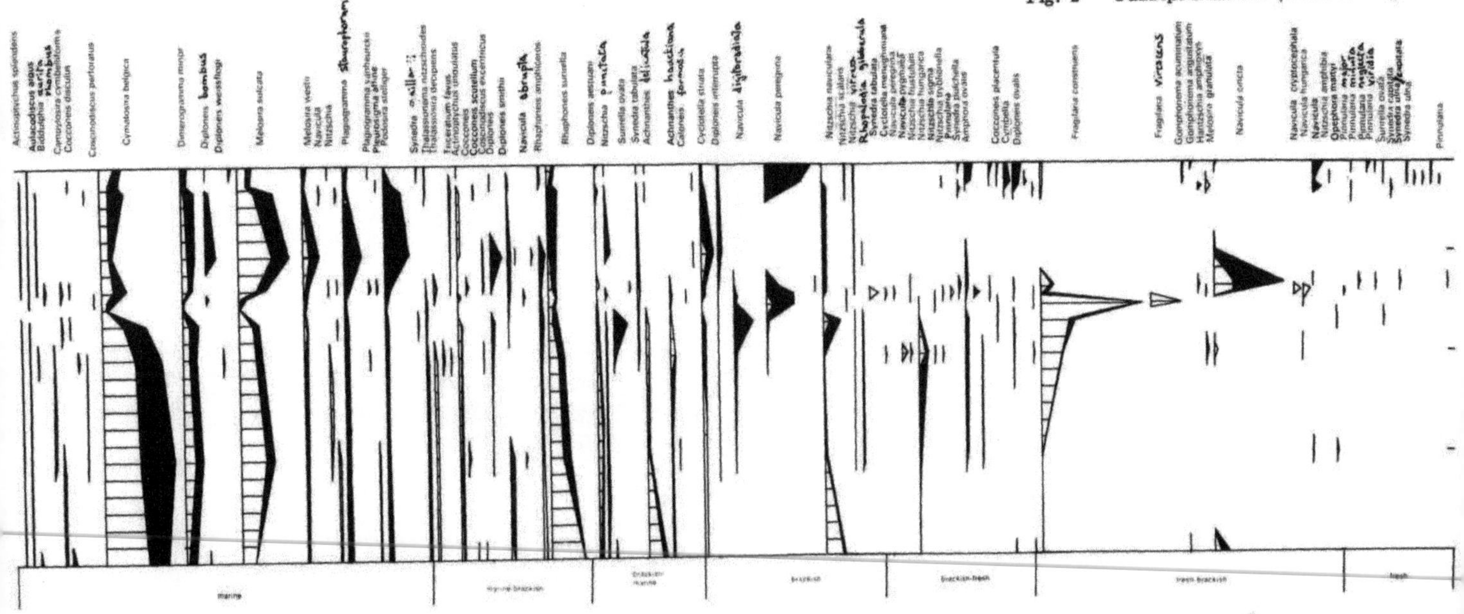

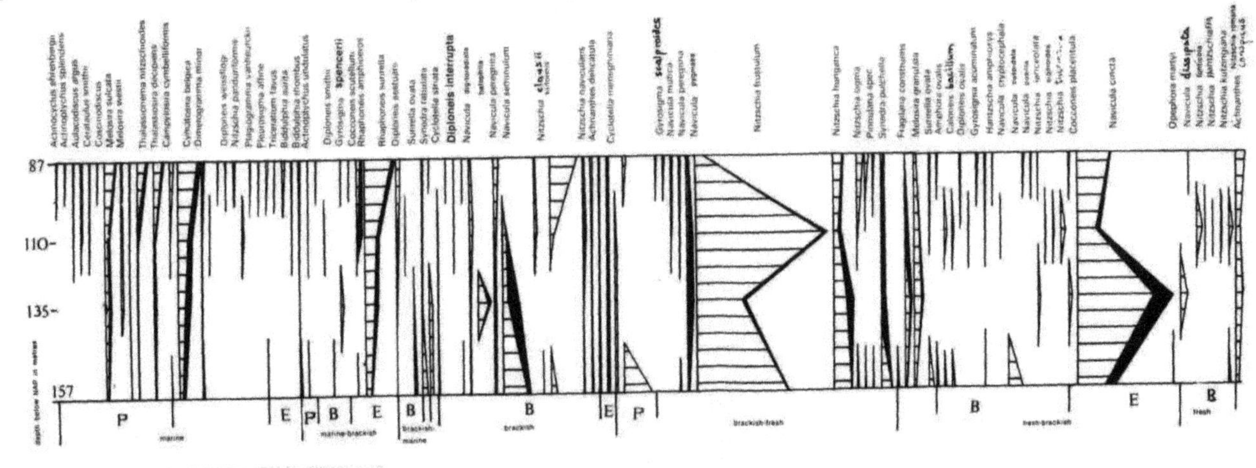

Fig. 3 Paddepoel III-314 (ditch diagram).

Please note that a full-size version of the figures on this page is available to download from www.barpublishing.com/additional-downloads.html
The original foldout has been reduced in size to match the A4 format of this book, the images are therefore not as clear as the original foldout. Please refer to the original foldout via the download for the original content.

can be seen on the following diagrams, one from an enclosure ditch, the other from a field ditch (figs. 4 and 5).

Differences of this order may be explained as follows: the original sediment, deposited in the Dunkirk I transgression, contained marine and marine-brackish diatom species, predominantly in damaged condition. The fresh-water diatoms actually lived in the ditches, and are, therefore, autochthonous. Why, then, are these individuals from the field boundaries also damaged? The answer seems to be that the damage is the result of periodic drying up of the ditches, which provides further proof that the fields were for arable use only. In view of their slightly higher position, the enclosure ditches must certainly also have fallen dry at times, but due to the rich organic fill (household rubbish) the diatoms are rather better preserved.

b. Diatom analysis of pottery and the correlation with sediment samples

In a number of cases, pottery from coastal settlements in the Netherlands has also been subjected to diatom analysis, in order to establish the degree of salinity of the water in which the clay utilized was originally deposited. Not every sherd will contain diatoms, by any means. Pottery may be made from a clay which never contained diatoms to begin with. In other cases, the pottery may have been fired at a temperature above 800°C, thus destroying any diatom shells which might have been present originally. By comparing the diatom content of pottery with that of the locally available clay deposits it is, under favourable circumstances, possible to establish whether the clay used in the manufacture can be of local origin or not. If the spectrum obtained does not match the local clays, either clay or pottery must have been brought from further afield. If they match, the clay may be obtained locally, but one is not sure that clay with a similar diatom content could not have come from elsewhere. In comparisons of this type it is necessary to determine.

1. whether the species in the clay of the sherd are the same as those in the local clay,

2. whether the proportions of individuals in each species are consistent in both samples and

3. whether the proportions of whole to damaged specimens are consistent for each species.

If one or more of these conditions cannot be met, then the clay must have come from elsewhere, in which case it is most likely that the pottery concerned was imported.

I would now like to describe a few settlements as examples of analyses seeking the correlation between pottery and locally available deposits and the conclusions to which these respective studies led.

For my first example, I have chosen Vlaardingen near Rotterdam (fig. 1). Here, around 2350 B.C. was situated a settlement of the eponymous Vlaardingen Culture, while much later, around 1940 B.C., long after the Vlaardingen people had departed, people of the maritime Bell Beaker phase established themselves on the same site (van Regteren Altena, Bakker et alii, 1962-1963). Here the difference between the diatom content of some Vlaardingen sherds and that of the stratigraphically differentiated and later sherds of the Bell

Fig. 4 Rijswijk (diagram of enclosure ditch).

Fig. 5 Rijswijk (diagram of boundary ditch).

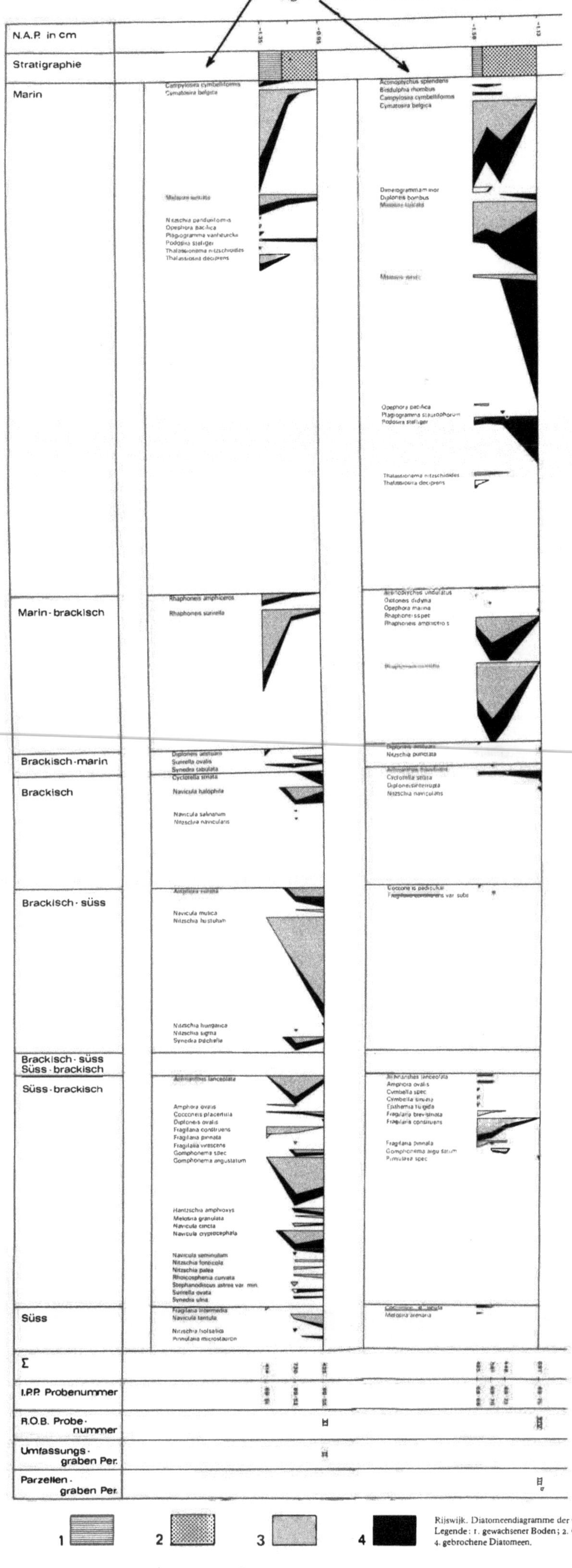

Rijswijk. Diatomeendiagramme der Gräben.
Legende: 1. gewachsener Boden; 2. Grabenfüllung; 3. ungebrochene Diatomeen; 4. gebrochene Diatomeen.

Please note that a full-size version of the figures on this page is available to download from www.barpublishing.com/additional-downloads.html. The original foldout has been reduced in size to match the A4 format of this book, the images are therefore not as clear as the original foldout. Please refer to the original foldout via the download for the original content.

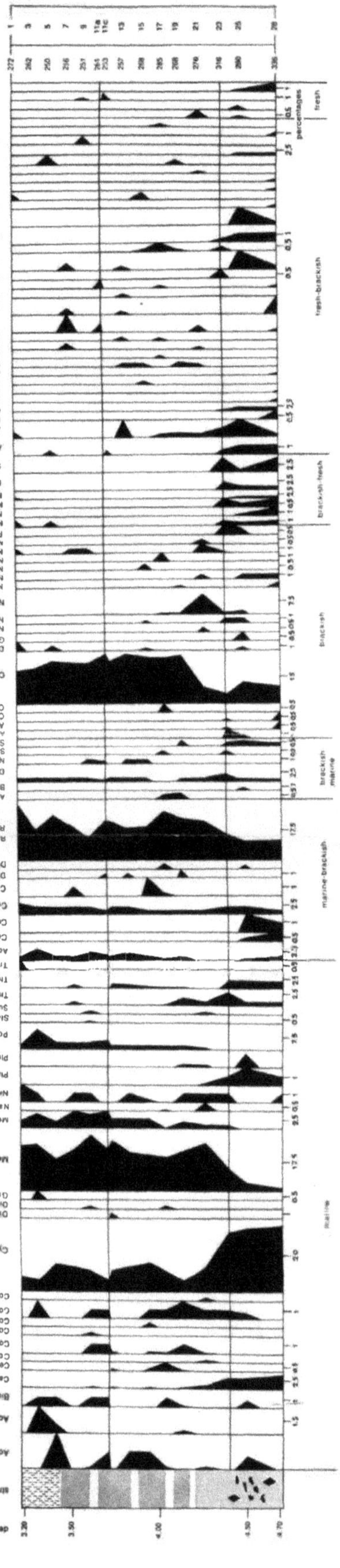

Fig. 6 Vlaardingen (creek diagram).

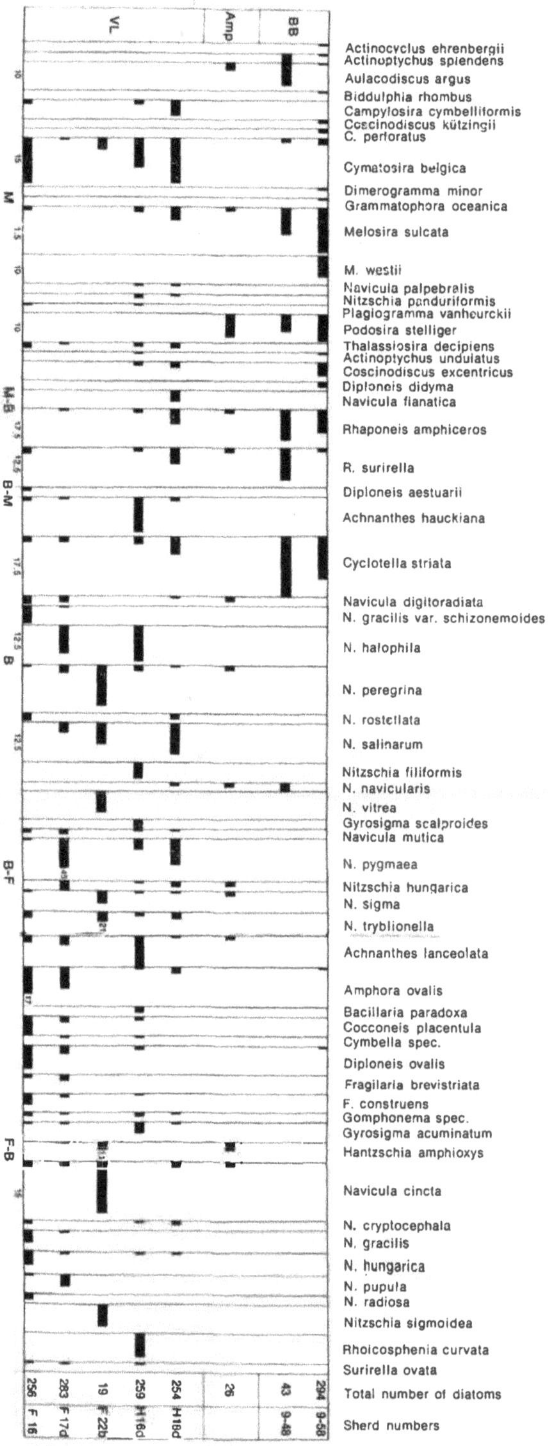

Fig. 7 Vlaardingen (sherd diagram).

Please note that a full-size version of the figures on this page is available to download from www.barpublishing.com/additional-downloads.html
The original foldout has been reduced in size to match the A4 format of this book, the images are therefore not as clear as the original foldout. Please refer to the original foldout via the download for the original content.

Beaker culture (maritime phase) was associated with a comparable change in diatom content of samples taken from a section cut on the site. It appeared that the diatom content of the Vlaardingen sherds correspond to that in the fill of a creek bed in the site, in which brackish to fresh water must have flowed. The sherds of the Bell Beaker culture, on the contrary, contained mainly allochthonous marine and brackish species, which again correspond to the content in the clay deposited just prior to and during the Bell Beaker habitation by the periodic flooding and breaking of the banks of the same creek, which had, in the meantime, displaced itself laterally (Groenman-van Waateringe & Jansma, 1969). Figure 6 will show a diagram of the section, while figure 7 will show sherds of the Vlaardingen and the Bell Beaker culture.

I will draw my second example from the Iron Age settlement at Middelstum in the province of Groningen (fig. 1) (Boersm, 1970). Here three Ruinen-Wommels-I sherds were analysed, along with samples from a creek section and an old habitation surface (Jansma, 1977), In the diagram from the creek (fig. 8) the complete specimens are, as we can see, concentrated especially in the brackish to fresh range, but occur also amongst some marine and marine-brackish species. However, since these species have tough shells they are less indicative of the environment than, for example, the unbroken individuals of the long and hence easy breakable brackish-fresh species. In this case, slightly brackish, stagnant water might be visualized. A sample from the old habitation surface shows the diatoms present in the clay cover deposited there. A number of species occurring in the fill diagram are absent here, namely the autochthonous creek vegetation, that is the species from a brackish to fresh environment.

The analyses of two of the three sherds correspond to the creek clays, not to the clays on the old habitation surface (fig. 9). The third sherd contained only six diatoms. A similarly low count was obtained from three Ruinen-Wommels-I sherds from Hijkerveld, in the province of Drente (fig. 1) (Harsema, 1974). The diatom content of three Hijkerveld sherds—Hijkerveld being an inland site—and the two sherds from the cutting in the coastal site of Middelstum is perfectly consistent with the use of nearby clays. Why the third sherd from Middelstum should differ is unclear, since the fabric is identical to the other two sherds and it is also, to the eye at least, indistinguishable from the Hijkerveld sherds.

The analysis of some medieval sherds from Medemblik, province of North Holland, forms my third example (fig. 1) (Besteman, 1976; Jansma, 1977). Here it was also possible to take cores from a section in the shore zone of the former Lake Wervershoof (Voorrips & Jansma, 1974). As can be seen on fig. 10 four stratigraphic layers are distinguishable. To start at the bottom: a layer of clay deposited in fresh water is visible covering the scanty Calais IVB deposit (layer I). Here only fresh-brackish and fresh species occur. Broken individuals are nearly absent here, so it is justified to assume that these species lived on the spot, and there are indications that sedimentation in an environment with calm, fresh water took place. In the next layer (layer 2), a large number of marine, marine-brackish, brackish-marine, brackish and brackish-fresh diatoms appear, so it is possible that the environment was gradually silting up due to transgression processes; but it is also possible, because of the many broken individuals, that an inflow of disturbed material,

Fig. 8 Middelstum (creek diagram).

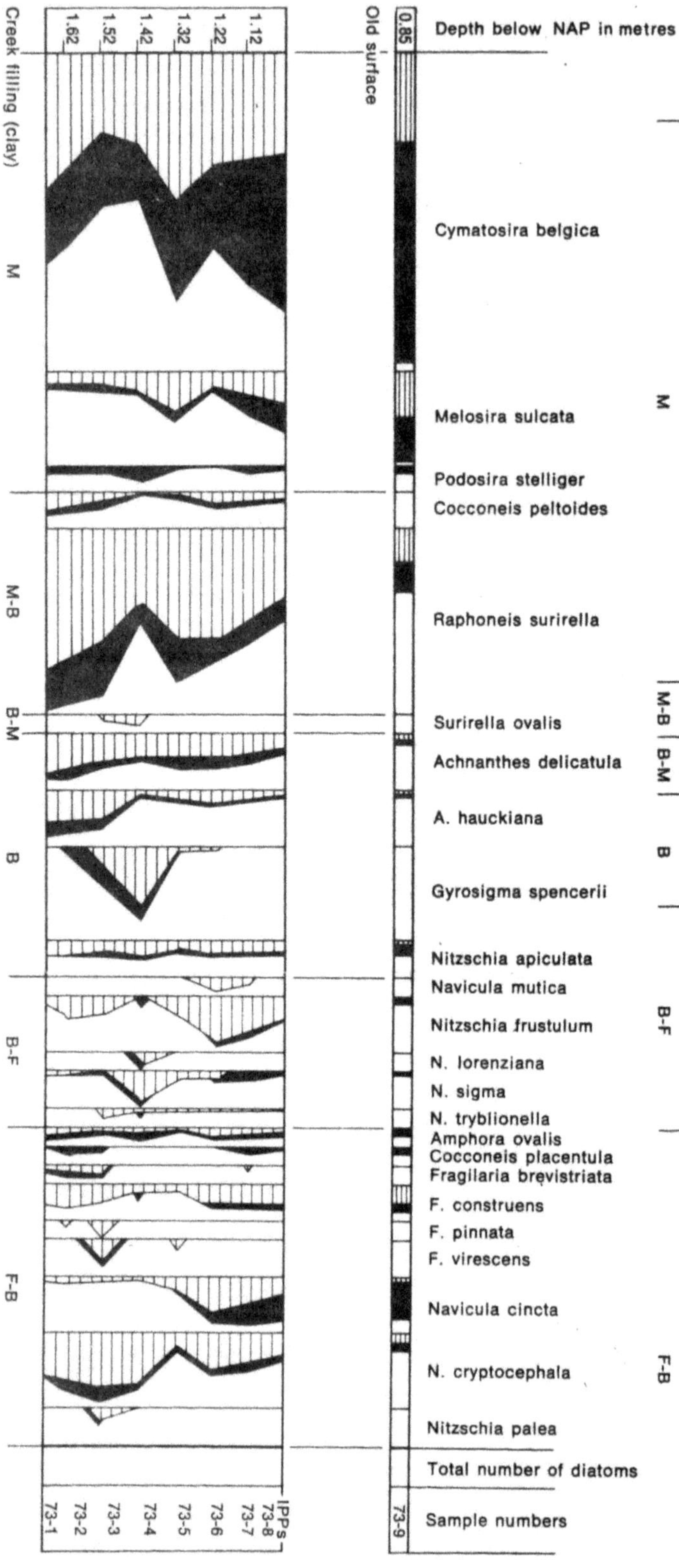

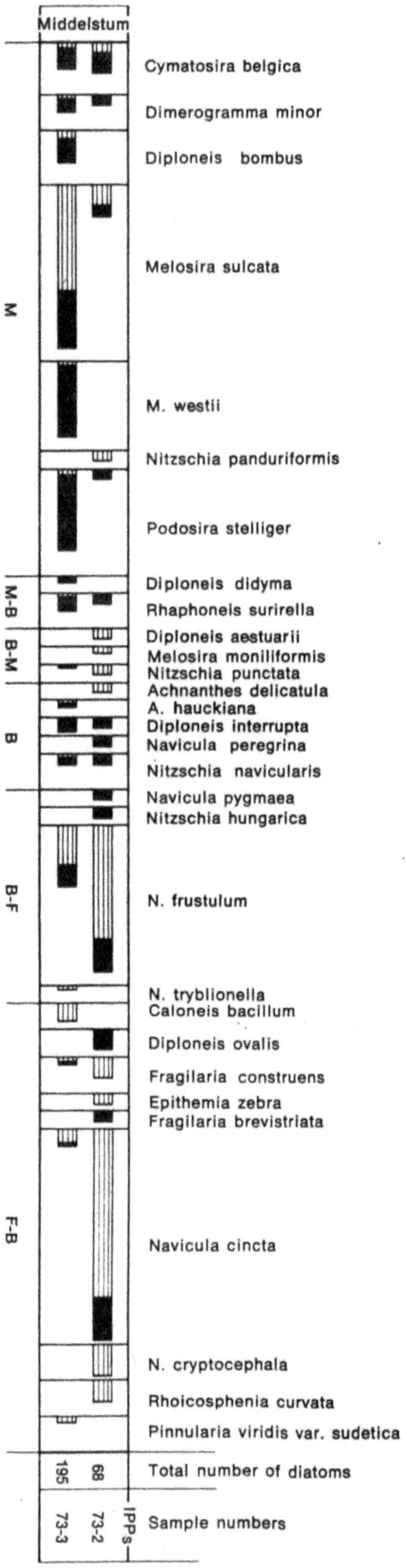

Fig. 9 Middelstum (sherd diagram).

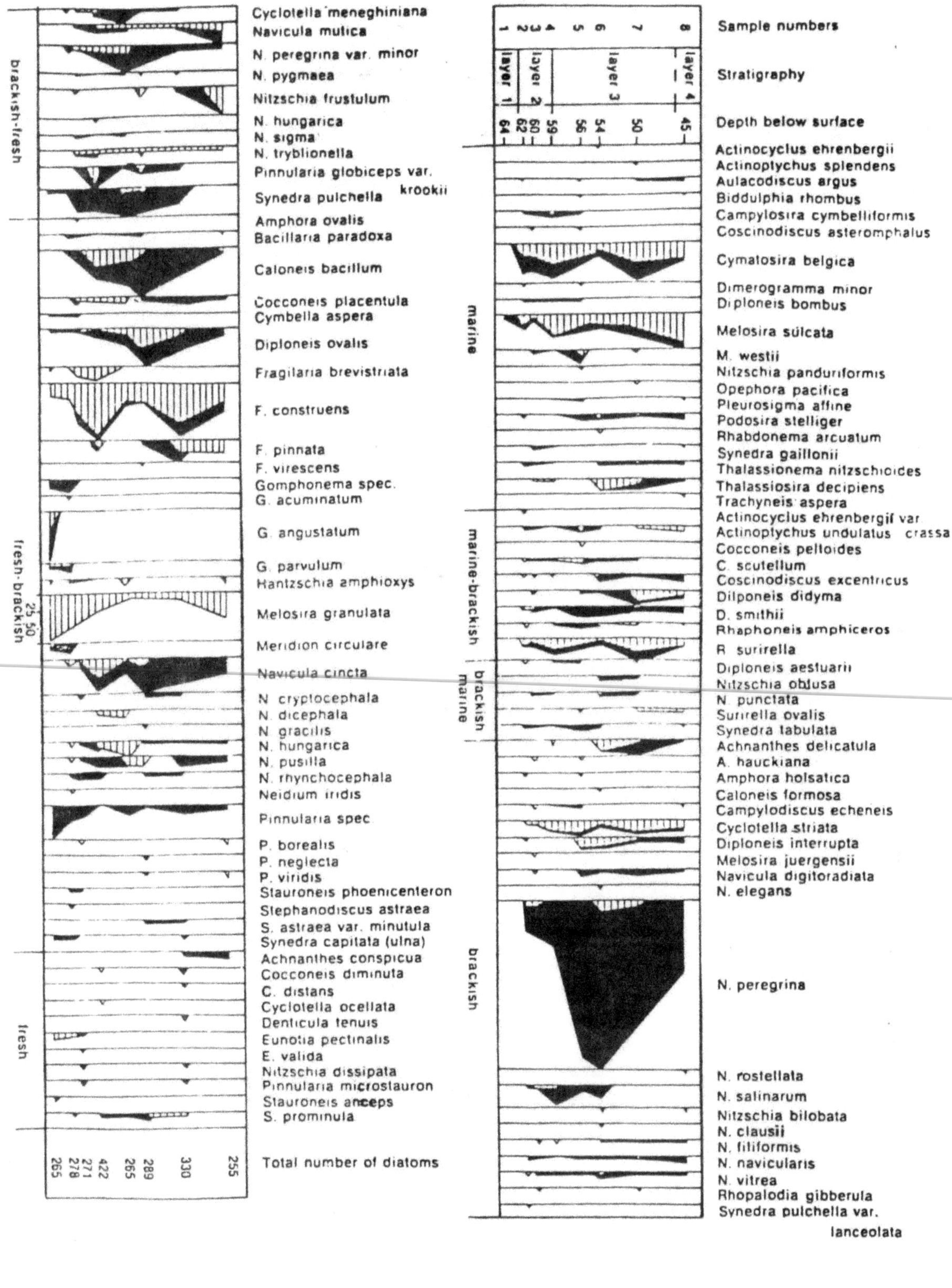

Fig. 10 Medemblik (diagram of section).

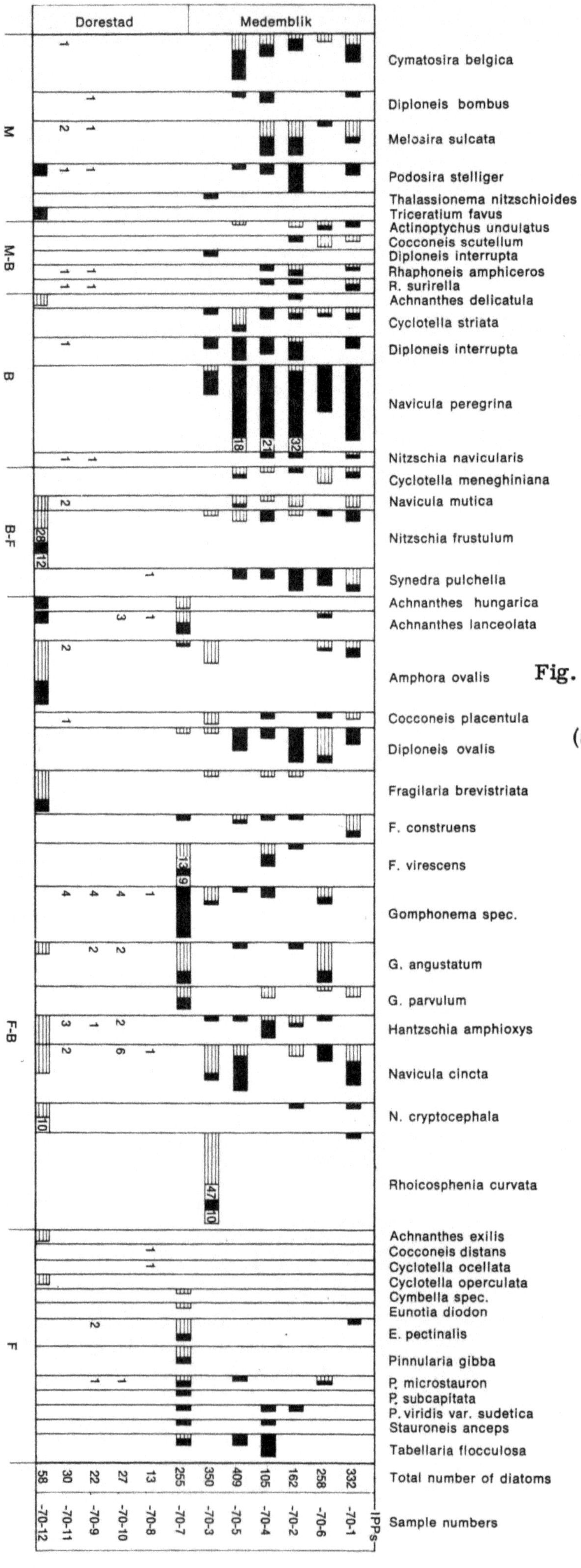

Fig. 11 Medemblik (sherd diagram).

originally deposited in a marine-brackish environment, took place, the material now being redeposited under predominantly fresh-water conditions. During the deposition of the "kiek" clay (layer 3), only slight changes in the environment are discernable: marine, marine-brackish, and brackish species remain, displaying roughly the same ratio of whole to broken specimens as in the previous layer. There is little change among the marine species in layer 4, a burnt layer which marks a Carolingian occupation. In the case of the marine-brackish types some show a decline in both broken and unbroken specimens, while others are represented only by broken ones. Considering the diagram as a whole, —perhaps with an exception for the diatoms from layer 1—it is difficult to form an opinion as to the true autochthonous diatom flora, if it was present at all, since the proportions of complete to broken individuals are almost identical for all species: marine, brackish and fresh. The general impression obtained is, that most of the diatoms in the "kiek" clay (layer 3) are secondarily redeposited. A few sherds of "Kugeltöpfe" were also examined for diatoms (fig. 11). The sherds analysed are 12 in all. As we see, the numbers I.P.P.S. 70-1 to 6 are from Medemblik, the numbers 7 to 12 from Dorestad, a medieval post about 10 miles SE from Utrecht (fig. 1). The Medemblik sherds no. 1 and 6 are from period I, that means the time in which the earliest Carolingian "kiek" clay was deposited, the sherds from Medemblik nos. 2, 4 and 5 are from period II, the so-called "daub"—level, while the Medemblik sherd number 3 has to be dated to the 12-13th century. There was a great measure of similarity between the diatom content of these sherds and the spectrum from the "kiek" clay deposit obtained from the aforementioned section. In comparing the diagram for the sherds with the diagram for the section we have got the impression, that the clay for pottery making was taken from the superficial "kiek" clay deposits available locally (layer 3). As far as the diatom content is concerned there is no difference between sherds from period I and II. An exception is the much later 12-13th century Kugeltopf, sherd number 3, since marine, marine-brackish and brackish-marine species are almost entirely absent while the valves of the species in the brackish-fresh to fresh-brackish range are predominantly complete. In addition, the absence of a peak for the brackish species <u>Navicula peregrina</u> and the large number of the fresh-brackish species <u>Rhoicosphenia curvata</u> are noteworthy. According to B. J. Cholnoky (1968) this latter species has a preference for running water in a more or less alkaline environment. So we can ask ourselves if perhaps clay from a brackish-fresh stream was used in this case. As mentioned above, the sherds nos 7-12 are from Dorestad. Number 7 is grit tempered and contains almost exclusively fresh-brackish to fresh diatoms and the same appears to be true for the other grit tempered sherds, the nos 8 and 10. On the other hand, the sherds with shell temper, nos 9 and 11, contain a very few marine to brackish diatoms, so few, in fact, that manufacture in a coastal region or the import of coastal clays need not to be considered and the diatoms may have come with the shell temper, which presumably was brought from the coast. The few marine diatoms present in sherd no. 12, which was grog tempered, may similarly be explained by the presence of diatoms in the grog temper. The diatom spectrum of this sherd diverges from the others, chiefly in the presence of the brackish-fresh species, which not only occurs in considerable numbers but also displays a high ratio of unbroken shells. Concluding we may say, that the analysis

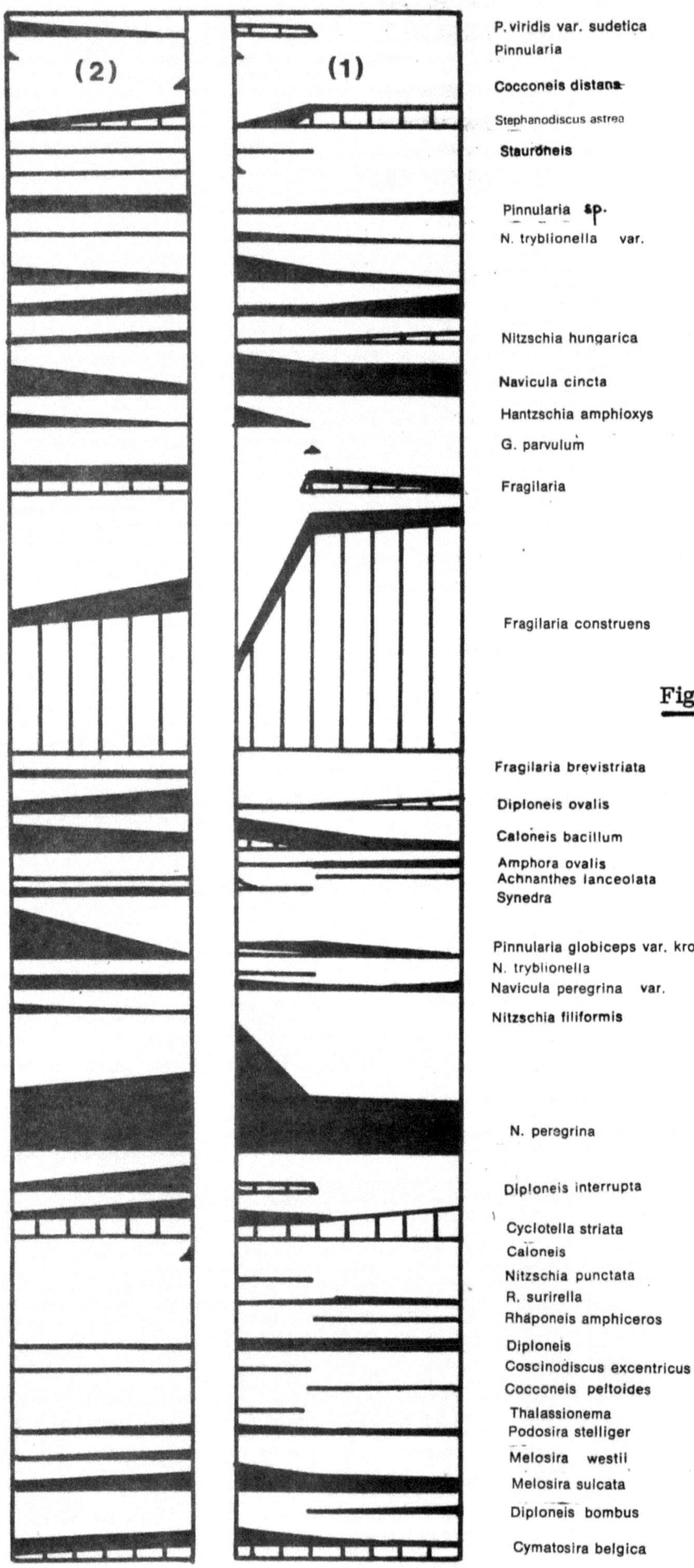

Fig. 12 Haarlem

suggests a local production of Medieval Kugeltöpfe, that is, if such a conclusion is warranted from the analysis of material from only two sites.

That the clay for pottery manufacture was not invariably extracted from the nearest available source can be illustrated by an example from the town Haarlem, some 15 miles west of Amsterdam (fig. 1). Sometime around the 12th century A.D. a potter's kiln was situated on the southernmost edge of the town centre. For the 12th century potter there were two sites with easily extractable clay open to exploitation. One site lay about 1 mile to the NE, near Bakenes within the old town limits, while the other lay approximately 5 miles further on, between Haarlem and the North Sea Canal in the Velserbroek Polder. The clay from Bakenes was somewhat humic and contained predominantly fresh water diatoms, while that from the Velserbroek Polder must have been deposited under rather more saline conditions in view of the diatom content. In all cases, marine and marine brackish species dominated, although the total numbers of individuals per sherd were quite different from each other. Comparison with the diatom spectrum from Bakenes on the one hand and that from the Velserbroek Polder on the other, showed that the diatom content of the sherds is best matched in the Velserbroek Polder, especially in its marine species, while the number of individuals per species and the ratio of whole to broken specimens is also consistent, as can be seen on fig. 12.

Although it is not strictly justified to apply the conclusion drawn from the examination of only a few sherds to all the product of the kiln, it does look as though clay was brought from the more distant Velserbroek Polder in preference to that from Bakenes, possibly because it was more suitable for the products of the kiln.

III. Diatom analysis from coastal sites abroad

Diatom analysis of pottery has until now only found application in the Netherlands. I was therefore delighted to receive four sherds from Mrs. Margaret Jones from Mucking, Essex (fig. 13) (Jones, 1968 and 1969). The sherds come from early Saxon pottery from the 5th century, recovered from floors, ditches and pits. The first sherd examined is from a bowl ornamented on the carination with the continuous faceting which is—according to J. N. L. Myres—characteristic of the early 5th century. But it also carries a feature typical of the earlier Germanic pottery, which survived into the 4th century in England on Romano-Saxon wares. Perhaps there are close parallels to the pottery from Feddersen Wierde. Anyway, the sherd contained very few diatoms (I could count only 6 specimens), all living in fresh water in a eutrophic and alkaline environment. The second sherd also dates from the 5th century and belongs to coarse ware. This contained only a few tiny diatom fragments. so small that further determination was impossible, with the exception of a single fragment attributable to the Pinnularia family, a predominantly fresh water species. The third sherd, from a pot from a Post-Roman enclosure ditch fill, also contained only a few individuals (total 6), only one of which was broken. The remaining individuals all belong to species inhabiting predominantly fresh and possibly even stagnant water, though they also possess a high toleration of salt water. Finally, the fourth sherd also comes from a pot in a Post-Roman enclosure ditch fill. Seven diatom fragments from this sherd also point to the deposition of the clay in fresh water conditions.

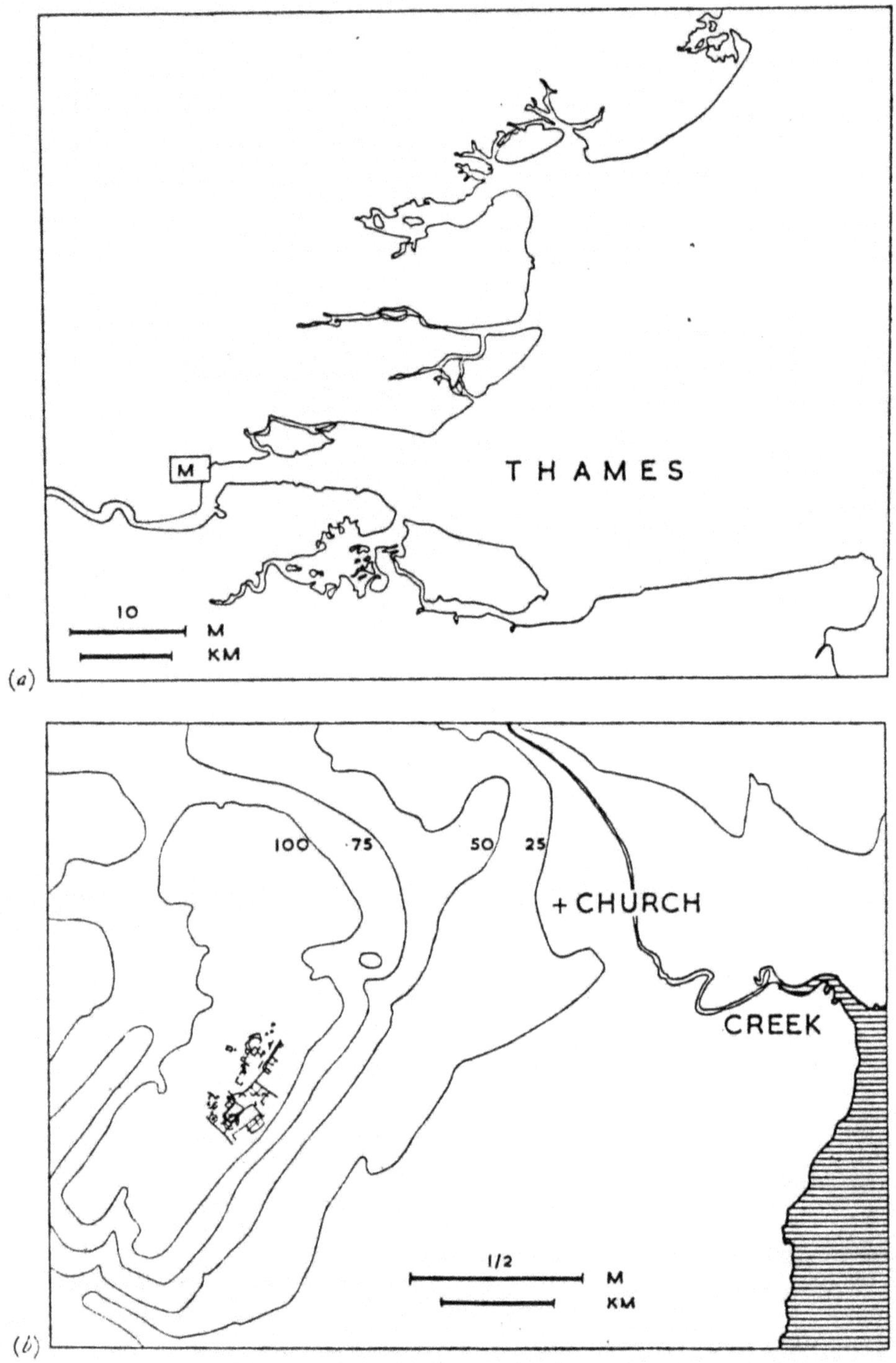

Fig. 13 The site of Mucking, Essex. Upper: general location with site marked M. Lower: Mucking area with crop-mark sites. Modified from M. Jones (1974) *Antiq. J.*, 54, p. 184.

Only after analysis of the locally occurring sediments can any conclusions be drawn with regard to local production versus import.

IV. Conclusions

In the preceding pages I have tried to give you an impression of the extent to which diatom analysis may assist in reconstructing the environment around an ancient settlement and how the analysis of pottery can lead to a better insight into the processes of manufacture; whether the pottery had been made locally from local materials, or whether there was import, either of clay or of the pottery itself. In both cases, the investigation may have implications for the trading contacts of the community.

Note:

Translated into English by Dr. Carol van Driel-Murray

BIBLIOGRAPHY

Altena, J. F. van Regeteren, Bakker, J. A., Clason, A. T., Glasbergen, W., Groenman-van Waateringe, W. and Pons, L. J., 1962-1963. The Vlaardingen Culture. Helinium 2, 1962, 3-35, 97-103, 215-243; 3, 1963, 39-54, 97-120.

Besteman, J. C., 1976. Carolingian Medemblik. Ber. R. O. B. 24; 43-106.

Bloemers, J. H. F., 1978. Rijswijk (Z.H.), "De Bult", Eine Siedlung der Cananefaten. Nederlandse Oudheden 8, 1978.

Boersma, J. W., 1970. Boerdam. Nieuwsbull, KNOB 12: 147-148.

Cholnoky, B. J., 1968. Die Ökologie der Diatomeen in Binnengewässern. Lehre.

Groenman-van Waateringe, W., 1978. Palynologische Untersuchung dreier Grabenprofile aus der römerzeitlichen Siedlung. In Rijswijk (Z.H.). Kapitel L aus: J. H. F. Bloemers, Rijswijk (Z.H.), "De Bult", Eine Siedlung der Cananefaten, Nederlandse Oudheden 8: 452-456.

Groenman-van Waateringe, W. and Jansma, M. J., 1968. Diatom analysis. Appendix III to W. A. van Es, Paddepoel, excavations of frustrated terps, 200 B.C. - 250 A.D. Palaeohistoria 14: 286.

Groenman-van Waateringe, W. and Jansma, M. J., 1969. Diatom and pollen analysis of the Vlaardingen Creek. A revised interpretation. Helinium 9: 105-117.

Harsema, O. H., 1974. Archeologisch onderzoek op het Hijkerveld, gem. Beilen. Voorlopig bericht van de campagnes 1969 en 1970. Nieuwe Drentse Volksalmanak 91: 161-168.

Hustedt, F., 1930. Bacillariophyta (Diatomeae). In A. Pascher, Die Süsswasserflora Mitteleuropas, Heft 10. Jena.

Hustedt, F., 1930-1966. Die Kieselalgen Deutschlands, Österreichs und der Schweiz unter Berücksichtigung der übrigen Länder Europas sowie der angrenzende Meeresgebiete. In L. Rabenhorst, Kryptogamen-Flora von Deutschland, Österreich und der Schweiz, Bd. 7. Tl. 1. Lepizig, 1930; Autorisierter Nachdruck, New York-Weinheim, 1962; Tle 2-3. Leipzig, 1959-1966.

Jansma, M. J., 1977. Diatom analysis of pottery. In B. L. van Beek, R. W. Brandt and W. Growenman-van Waateringe (eds.), Ex horreo: I.P.P. 1951-1976, Cingula 4, pp. 77-85. University of Amsterdam.

Jansma, M. J., 1978. Diatomeenanalysen einiger Grabenprofile in der römischen Siedlung von Rijswijk (Z.H.). Kapitel L aus: J. H. F. Bloemers, Rijswijk (Z. H.), "De Bult", Eine Siedlung der Cananefaten, Nederlandse Oudheden 8: 447-451.

Jones, M. U., 1968. Crop mark sites at Mucking, Essex. Antiq. J. XLVIII: 210-230.

Jones, M. U., 1969. Saxon pottery from a hut at Mucking, Essex. Berichten R.O.B. 19: 145-156.

Voorrips, A. and Jansma, M. J., 1974. Pollen and diatom analysis of a shore section of the former Lake Wervershoof. Geologie en Mijnbouw 53: 429-435.

Werff, A. van der, and Huls, H., 1957-1974. Diatomeeënflora van Nederland. Abcoude.

MAMMALS AND MAN ON ORONSAY: SOME PRELIMINARY HYPOTHESES CONCERNING MESOLITHIC ECOLOGY IN THE INNER HEBRIDES

By Caroline Grigson

1. INTRODUCTION

The Mesolithic occupation of the island of Oronsay in the Inner Hebrides is represented today by the remains of five known shell middens (see figure 1). They are all situated on ancient shorelines just above the level of the main postglacial marine transgression when the sea level was roughly 7 m higher than it is today, and they date from the final phases of the Atlantic Period. Radiocarbon dates ranging from 3,695 $\pm$ 80 b.c. to 3,200 $\pm$ 380 b.c. have been obtained from charcoal in the middens (Mellars 1978).

Caisteal-nan-Gillean I was first investigated in the 19th century (Grieve 1885, Anderson 1898) and Cnoc Sligeach at the turn of this century (Anderson 1898, Bishop 1914); the results of these excavations are conveniently summarized by Lacaille (1954). The Priory Midden, Caisteal-nan-Gillean II, and Cnoc Coig were all found during the course of investigations by Dr. Paul Mellars and his colleagues between 1970 and 1979. Large scale excavation of Cnoc Coig has just been completed and Caisteal-nan-Gillean I and II, Cnoc Sligeach and the Priory Midden have been sampled; a preliminary account has been published by Mellars (1978); various specialists are involved in the investigations of the middens and their contents and it will be some time before the work is finished and the final results appear.

This essay is an interim stage in the study of the mammalian remains from the new excavations, which, together with the material already published, allows the formulation of various hypotheses. Some of these questions have already been discussed by Mellars (1978), but this is an attempt to further selected lines of argument in the light of the new mammal identifications and a more intensive study of Oronsay and the neighbouring island of Colonsay as they exist today. Eventually these hypotheses will be tested on the basis of a detailed analysis of the mammal bones and an attempt will be made to reconstruct as far as possible the natural and human ecology of the time. In order to do this the mammal bones will not, of course, be studied in isolation, but in relation to the other, copious animal remains, to the location of the sites, and to the early geological, botanical and faunal history of the islands. In the present study these aspects will only be touched on where they are directly relevant to the hypotheses being advanced.

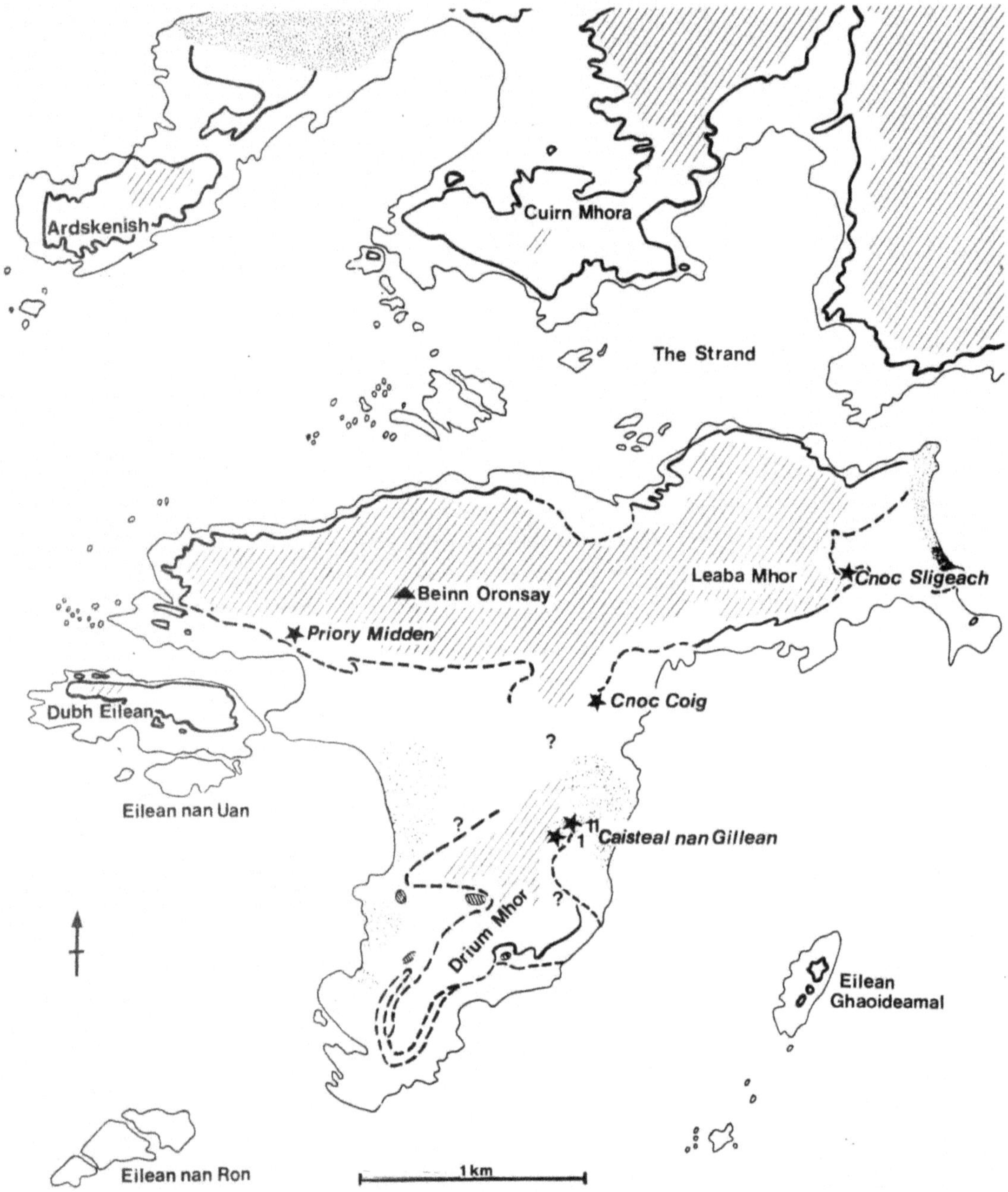

Figure 1. ORONSAY AND THE SOUTHERN PART OF COLONSAY

Showing how the Mesolithic shell middens ★ are situated on the main Holocene shoreline, marked --- (from Jardine 1978). The approximate position of the rest of the Holocene shoreline is represented by the present 10m contour —— .

The coasts of both islands were more indented in the Mesolithic than at present, and the southern part of Oronsay may have been a separate island. Of the two modern grey-seal breeding islets Eilean nan Ron would have been submerged at high tide and Eilean Ghaoideamal nearly so. It is suggested that seals may have bred on the east side of Oronsay during the Mesolithic.

/ mean high water springs; land over 15m; modern sandhills; modern lochrans

2. MODERN CONDITIONS

2.1 Location

Oronsay and Colonsay are islands situated in the mouth of the Firth of Lorne, 16 kilometres from the large island of Mull to the north, 9.5 km from Islay to the south and about 12.5 kilometres from Jura in the east (see figure 2). The entire southern coast of Mull and much of both Islay and Jura are visible from Colonsay and Oronsay, although the visibility varies greatly with the weather. At times almost every detail of the coasts and mountains of Islay and Jura stand out with crystalline clarity, at others they are blotted out by driving rain or mist. In very clear weather the hills of Donegal in Ireland may be visible from the top of Beinn Oronsay as a smudge on the southern horizon.

Oronsay is connected to Colonsay by a wide, sandy strand that can be crossed for about three hours around low water. Oronsay is a small island, triangular in shape; its northern land mass is about 4.5 km long and from the north coast to the southern point is about 2.5 km. Its area is approximately 550 hectares. Colonsay is much larger: 13.5 km along its main axis (from SW to NE) and varying in width from about 2½ to 5 km. Its area is about 4000 hectares.

2.2 Present landscape

Both islands are formed of lower Torridonian sedimentary deposits which have been folded and weathered into a miniature highland landscape of rocky mountains and boggy valleys, many containing small lakes or lochans. The mountains are quite low in terms of absolute elevation: the highest point on Oronsay is Beinn Oronsay at 93 m, and on Colonsay the highest point is Carn an Eoin at 143 m.

The northern part of Oronsay is fairly high and rocky with Beinn Oronsay to the west and an area of low rocky hills and boggy valleys to the east. At the far eastern tip there is an ancient beach sloping towards the sea, but separated from it by high coastal dunes. The southern part of the island is low-lying and consists largely of raised beach deposits of both pebbles and sand, with a wide area of grassy machair, some dunes, a few rocky outcrops and some marsh. The interrelationships between the modern and ancient dunes and the middens are very complicated and are to be studied by Dr. D. Gilbertson. Several radiocarbon dates on shells from within the pebble beaches of Oronsay have been obtained by Dr. Graham Jardine, who had done much research on the correlations between the raised beaches, the middens and ancient sea levels (Jardine 1977 & 1978).

Although there are dramatic sea cliffs on the north-western shore of Colonsay below Beinn Bhreac and Beinn Riabhach, most of the coast line of both the islands is fairly low, and the rocky headlands are interspersed with sand and pebble beaches both at sea level and at heights of up to about 11 m, and occasionally, even about 27 m O.D. As the rock surface of Oronsay and Colonsay was scoured clean of all loose material by the last main advance of the Devensian ice sheet, these raised beaches must have been formed since the retreat of the ice. The highest probably represent storm beaches thrown up during the Late Glacial period (Cunningham Craig et al. 1911).

Figure 2. THE POSITION OR ORONSAY AND COLONSAY IN RELATION TO ISLAY AND JURA

Showing some of the features mentioned in the text, and the sea depths (in fathoms) around the islands (1 fathom = 1.8 metres). The shallowest water lies between Oronsay and Islay, off Ardnave Point.

The mean low water spring tide level is shown by the dashed line; note that Oronsay is joined to Colonsay at low water.

2.3 Flora

The present flora of the islands consists typically of bog in the upland valleys, with heather on the drier slopes, and grassland and bracken on the coasts and in the better drained inland valleys. However, this open, treeless appearance is rather deceptive: there are rowan trees *Sorbus aucuparia* and hawthorn *Crataegus monogyna* scattered about in most parts that are sufficiently sheltered from the south-west wind, and small species of willow (*Salix*) abound in marshy areas. On Colonsay in very sheltered places many deciduous and some evergreen trees planted in the eighteenth and nineteenth centuries still thrive.

There are as well two small, dense woods which may well be relicts of more widespread natural woodland. These are Coille Mhor and Coille Bige on, not surprisingly, the most sheltered coast, the north-eastern (figure .2). They consist largely of birch *Betula* sp., small oak trees *Quercus* sp., rowan trees, and, importantly, hazel nut trees *Corylus avellana.* McNeill also noted aspen *Populus tremula,* hawthorn and holly *Ilex aquifolium.* I am indebted to Dr. Gilbertson for the observation that birch is regenerating in the ungrazed valley to the south of Coille Mhor.

Oronsay has no woodland—as Frances Murray wrote about the island in 1887, "with the exception of one small rowan tree, there is not so much wood to be had as would make a peastake or a porridge-spurtle", (quoted by Loder 1935); but even on Oronsay small willows abound and McNeill also noted some elder trees *Sambucus nigra.* Dr. Gilbertson recently noticed two very stunted oak trees and some rowan trees clinging to the lee side of Beinn Oronsay. Without in any way pre-empting the forthcoming palaeobotanical studies of the islands one can quote evidence for more extended woodland on Colonsay in the past: bog oak and goat willow stumps under the water of Loch Fada (an inland lake) and hazel nuts deep in the adjacent peat (NcNeill 1910). There is evidence of early post-glacial woodland on some western coasts and islands of Britain that are now bare of trees (Spencer 1975) and on some other islands — Skye (Birks 1973) and Harris (Evans 1971); so it seems likely that open woodland would have been present on the more sheltered parts of Colonsay and probably on parts of Oronsay as well. The same conclusion has been suggested by Dr. C. Paul from his studies of the assemblages of land molluscs from the sand dunes underlying the middens (Paul 1976). However as Birks (1977) has emphasized, it seems unlikely that any of the more exposed western islands of Scotland would have supported complete woodland cover at any time during the postglacial, so one can envisage a landscape of variegated relief, covered with open vegetation interspersed with woodland—a congenial habitat for many animal species.

2.4 Fauna

The most striking feature of the terrestrial vertebrate fauna of Oronsay and Colonsay is the absence of so many species that are, or were until recently, common on the mainland. According to Loder (1935) there are no amphibians or reptiles, no insectivores, no voles, stoats (*Mustela erminea*), weasels *M. nivalis*), wild cats, badgers, foxes, deer or black rats. The polecat (*M. putorius*) is extinct on Colonsay as in the rest of Scotland, but was said to be plentiful in 1549, and there is a place on Oronsay called polecat hollow

(Làg nam Feòcullan). The pine marten (Martes martes) is also missing now, but there are several Gaelic marten place names on Colonsay (such as Leana an Taghain at Baleruminmore). The marten is arboreal in habit so this may be a further indication of greater tree cover in the not too distant past.

Rabbits are almost ubiquitous on both islands and cause a great deal of damage, undermining the machair and competing successfully with sheep for grazing. Even the spread of myxomatosis has had little effect. Rabbits were probably introduced from Barra in the eighteenth century and a warren was dug for them near Kiloran (McNeill 1910). It should be noted that the Gaelic words for rabbit Coinneann, and for warren Choineigh, occur in some place names on Colonsay and Oronsay and presumably postdate the introduction of rabbits to the islands, so the mention of animals in Gaelic place names is no guarantee of their antiquity. The common hare (Lepus capensis) seems to have come and gone, and has perhaps been introduced more than once (McNeill 1910).

Not surprisingly house mice (Mus musculus) and brown rats (Rattus norvegicus) are common and must have been introduced. The only other rodent is the field mouse (Apodemus sylvaticus), which may well have been accidentally imported by man (Corbet 1961). It is first recorded on mainland Britain in the Boreal or Atlantic period (Sutcliffe and Kowalski 1976).

The only carnivores on the islands are the semi-marine otters (Lutra lutra): they are said to be common along the strand between the islands, on the coasts of Colonsay below Baleruminmore and Ardskenish, and below the north-western cliffs, feeding on fish, seabirds and waders.

Common seals (Phoca vitulina) are sometimes seen hauled out on the skerries between Colonsay and Oronsay, but there are no suitable breeding grounds for them in the vicinity.

The islets of Eilean nan Ron and Eilean Ghaoideamal (see figure .1) support notable colonies of the grey, or Atlantic seal (Halichoerus grypus). The breeding colonies are not large: the maximum number of adults counted on both islets together on one day in September 1956 was 269, and the total number of pups born in a whole season has been estimated to be between 350 and 400 (Hewer and Backhouse 1960). As well as breeding on the islets in the autumn grey seals haul out at other times of the year on the skerries between the islands, on isolated rocks on the south-west of Colonsay and, particularly when the tide is low, on the erosional platforms around Eilean nan Ron, when their crying and mooing can be heard over the sound of the surf on most of Oronsay itself.

Common rorquals (Balaenoptera physalus) are seen in the seas round about, and a 78-foot blue whale (B. musculus) was stranded near Balnahard in July 1916 (Loder 1935).

Some records of the vertebrates of Colonsay have been published in the Provisional Atlases of the Amphibia and Reptiles and of the Mammals of the British Isles (Arnold 1973 & 1978). These differ slightly from Loder's account in that the distribution maps, based on records of before 1960, include pygny shrews (Sorex minutus), as well as common lizards (Lacerta vivipara) and slow worms (Anguis fragilis) for Colonsay, However, since the maps exclude rabbits, otters and brown rats, which certainly are present, it is clear that more work on the recording of the terrestrial vertebrates of the islands is needed. Neither Loder nor Arnold mention the herds of feral goats that live in the Balnahard region and on Oronsay.

Colonsay and Oronsay have a rich and varied bird population of both residents and migrants which was listed by John Bevan and Lewis Palmer from 1930 to 1933 (printed in Loder, 1935). Fulmars (Fulmarus glacialis), kittiwakes (Rissa tridactyla), guillemots (Uria aalge), razorbills (Alca torca), and many species of gulls nest in huge numbers on the north-western sea cliffs. Puffins (Fratercula arctica) were seen occasionally, but do not breed. Eider ducks (Somateria mollissima) breed all over the islands and are commonly seen along the coasts; there are many other ducks and waders.

3. RED DEER

Red deer bones and antlers (or parts of antlers) are quite common in the middens of Cnoc Coig, Caisteal-nan-Gillean I and Cnoc Sligeach. The main questions that they raise seem to be: why are there red deer remains in the middens, that is, was use made of deer and where were the deer obtained from?

3.1 The use of red deer

Three main uses of deer (or parts of deer) suggest themselves:

1. The antlers and some bones were used for the manufacture of implements.
2. Venison and offal were eaten.
3. Deer skin was used (for clothing, shelter or boats).

There are of course many other possible uses for many parts of the body that do not survive in the archaeological record, such as sinews, guts, whiskers and so on. The various possible uses are not mutually exclusive and it may well be that deer were used for some, or all, of these purposes. This could best be tested by a bone element analysis:

a. If only antler (either shed or unshed) and bones used to make tools (usually metapodials) are represented on the site, it is obvious that the only use made of deer, at least at the middens themselves, was tool manufacture. In this case one would expect to find the waste products of manufacture as well as the tools themselves.

b. If only prime meat bones, that is vertebrae, scapulae, humeri, pelves, and the proximal ends of radii and tibiae (Uerpman 1973) are present it would suggest that the animals were butchered elsewhere and brought to the site for consumption.

c. If there is an overproportion of the bones of the skull and feet (which might be left inside hides) it would suggest that deerskin was being utilized.

d. If the bone elements are randomly represented it would seem that the animals were butchered at the site and put to a variety of uses. In this case it might be difficult to work out exactly what was being done with the carcases.

One very important proviso that applies to all bone element analysis is that of differential preservation, incorporation and retrieval of bones in an archaeological deposit. Binford and Bertram (1977) have shown that bone survival can be greatly affected by the action of scavenging dogs. So far dogs have not been safely identified from any of the Obanian sites, including those on Oronsay, so this may not be important. They also showed that less dense bones are more likely to be destroyed in the course of time. Factors which influence bone survival are complicated and not yet fully studied, but certainly include pH and water activity. The main midden material is obviously calcareous and its pH will probably be consistantly high, but leaching can occur and so samples of the middens will have to be assessed for variations in pH, and it may be possible to relate the results to the distribution of various bone elements within the middens. Differential retrieval undoubtedly occurred in the excavations of Symington Grieve and Henderson Bishop at Cnoc Sligeach and Caisteal-nan-Gillean I. Although much of the bone in Cnoc Coig was extremely fragile on exposure, very careful consolidation both *in situ* and in the laboratory by Jill Goudie and others has ensured the retrieval of almost every surviving bone. Sieving was varried out on samples selected on a probablistic basis (Peacock 1978) and greatly enhanced the collection of small bones, particularly those of birds and fish. The retrieval of two left scaphoids (tiny wrist bones) of otter, giving a minimum number of two individual otters, shows how effective the Cnoc Coig excavation has been. A secondary problem that can affect bone element analysis is that certain bones are more difficult to assign to species than others, particularly if they are incomplete.

3.2 The source of the Oronsay red deer bones and the zoogeography of Oronsay and Colonsay

Where did the deer represented in the Oronsay midden live? On Colonsay, or Oronsay itself, or on the mainland (or rather on Islay or Jura)? Colonsay, with an area at that time of about 3,400 hectares, would certainly be a possibility. According to Mitchell *et al.* (1977) the minimum area of woodland required to maintain a herd or red deer in Europe today is about 1,000 hectares, but the areas used today in Scotland can be much less, particularly on islands (Mitchell, p.c.). It is suggested in section 2.3 that Colonsay was likely to have been quite largely wooded and the combination of open woodland and open ground with varied boundary zones between the two would according to evidence quoted by Simmons (1975) be very suitable for high population densities of red deer. It seems (Mitchell *et al.* 1977) that what red deer particularly require is winter shelter, and the combination of trees and a landscape of varied relief, including many sheltered slopes and valleys might have been ideal. Oronsay on its own seems unlikely to have been able to support a breeding population of red deer as its area was only 300-400 hectares at the time of the maximum marine transgression, but deer might well have crossed over from Colonsay on occasion: as a modern parallel Dr. Brian Mitchell (p.c.) has records of deer swimming for distances of up to one kilometre to and from the isalnd of Scarba (near the northern tip of Jura).

On the other hand if the deer were caught on Islay or Jura it would seem likely that only young fawns (possibly alive) or antlers or parts of dead adult deer would have been transported by boat to Oronsay. Bone element analysis could help to elucidate this problem, but more positive evidence would be evinced if it could be shown that the deer population represented in the middens and hypothetically hunted on Colonsay differed anatomically from that of Islay and Jura. The most likely difference would be smaller body size in the Colonsay population: geographical isolation is known to lead to diminution in large ungulates. It could be argued that since Islay and Jura are also islands their red deer would be similarly affected; however both these islands are very large and would have been capable of supporting many herds of red deer, so its seems unlikely that the conditions for size reduction, whether genotypic or phenotypic, would have acted there, especially as the distances between Islay and Jura and between Jura and the mainland are so small that genetic interchange between the island populations and with the mainland deer would have been quite feasible. It is hoped that comparisons of size between the deer found in the Oronsay middens and contemporary deer remains from Jura and from the mainland will help to elucidate this problem.

If it were shown that the Colonsay deer were a separate population there would be the question of how they got there. Were they indigenous, or imported by man, or did they swim? To be indigenous they would have had to arrive in the area from the mainland at a time of extreme low sea level in early Holocene times. At present the shallowest waters (at about -22m O.D.) lie between the south of Oronsay and Ardnav point on Islay (see figure .2). Relative sea level in this district is thought to have stood at about +11 m O.D. between about 11,000 and 10,400 b.p. and to have dropped to around the present mean sea level at the beginning of the Holocene (10,000 b.p.), before returning to about +7 m at the time of the (local) maximum Holocene marine transgression. It seems likely therefore that, unless there has been a marked change in the submarine topography of the area, Colonsay and Oronsay have been islands since at least 11,000 b.p.

There is some indirect evidence for this view: a land bridge also seems unlikely on zoogeographical grounds, as so many of the small common British terrestrial vertebrates are absent from Colonsay and Oronsay (section 2.4 and table 1). The missing species even include some that are supposed to be late glacial relicts (common frog, arctic hare *Lepus timidus*, and stoat) and most of the remainder were present as well on the mainland in the early postglacial (Grigson, in press). The absence of some species from the islands today may be the result of quite recent extinctions (for example the polecat and the pine marten, mentioned above), but the table shows how restricted the small vertebrate fauna of the islands is.

It must however also be remembered that Islay and Jura (from which the bulk of Colonsay's fauna would have arrived if there had been a land connection) are also islands, albeit large in area, and also have a restricted fauna, though less so than that of Colonsay. However, although the channels separating them are very deep, Islay and Jura are very close to each other, and Jura is close to the mainland. The common animals which all four islands lack are the hedgehog (*Erinaceus europaeus*), the mole (*Talpa europaea*, red squirrel (*Sciurus vulgaris*), water vole (*Arvicola terrestis*), fox (*Vulpes vulpes*), pine marten (*Martes martes*), weasel (*Mustela nivalis*), wild cat (*Felis silvestris*) and badger (*Meles meles*); all are present on the nearby Scottish mainland (Arnold 1978) and most are known from the early postglacial as well (Grigson, in press).

TABLE 1 Terrestrial vertebrates on Colonsay compared with those of Islay and Jura

	Colonsay 1935 (Loder 1935)	Islay, or Jura, or both, since 1960 (Arnold 1973 & 1978)
Pygmy shrew (*Sorex minutus*)	-*	+
Common shrew (*S. araneus*)	-	+
Arctic hare (*Lepus timidus*)	-	+
Common hare (*L. capensis*)	-	+
Bank vole (*Clethrionomys glareolus*)	-	+
Short-tailed vole (*Microtus agrestis*)	-	+
Long-tailed field mouse (*Apodemus sylvaticus*)	+	+
Stoat (*Mustela erminea*)	-	+
Roe deer (*Capreolus capreolus*)	-	+
Red deer (*Cervus elaphus*)	-	+
Common frog (*Rana temporaria*)	-	+
Common toad (*Bufo bufo*)	-	+
Adder (*Vivipara berus*)	-	+!
Slow worm (*Anguis fragilis*)	-!	+
Common lizard (*Lacerta vivipara*)	-!	

It should be noted that the common otter (*Lutra lutra*) is not included in this list, as although it is common on Islay, Jura, Colonsay and Oronsay, it is semi-marine here.

* recorded as present between 1900 and 1959 (Arnold 1978);

! recorded as present before 1960 (Arnold 1973).

Table 1 emphasizes that the fauna of Oronsay and Colonsay lacks many species that exist on the neighbouring islands of Islay or Jura and suggests that on zoogeographical grounds Colonsay and Oronsay must have been separated from them for a very long time. If there were deer living on Colonsay they were not indigenous and must have either swum there or been imported by man.

Deer can swim quite long distances, but Oronsay is 9.5 km from the nearest land (Islay), the sea between them is swept by fast tides and includes. so it is said, whirl pools, It is true that this distance might have been reduced at a time when the sea level was lower than it is now, perhaps between the Late Glacial and Atlantic periods. Navigation seems unlikely to have been the high point of a deer's iintellectual capacity, but the possibility of the accidental arrival by sea of a pregnant hind or two cannot be ruled out.

It is quite possible that the deer were introduced by man: fawns, as Juliet Clutton-Brock (in press) has pointed out, can easily be trussed up and carried about in small boats. If Mesolithic man brought deer to the islands, could this be a first step towards controlled husbandry or even domestication? Small size in ungulates found in association with man is usually taken to indicate domestication, but only in the absence of other isolating mechanisms (Grigson 1969), so if the Oronsay deer do turn out to be of reduced stature, this could be accounted for by their geographical isolation. On behavioural grounds red deer are the most unlikely candiates for domestication, but the possibility of some sort of interference or control cannot be ruled out.

3.3 Red deer, auks and the possible use of Colonsay by Mesolithic people.

The main point to be learnt from this rather theoretical consideration of the implications of the presence of red deer bones in the Oronsay middens is that in considering the extent of the resource base of the inhabitants of Oronsay, Colonsay must be taken into account; a view which is reinforced by the study of the likely distribution of auk nesting sites within theislands.

Bones of the now extinct great auk (*Pinguinus impennis*) are quite numerous in the middens (Bramwell p.c. and Lacaille 1954). As these birds were pelagic, spending most of the year at sea, it seems likely that they would have been obtained when at their most concentrated and vulnerable, that is when nesting on land in Many, June and perhaps early July (Fisher & Lockley 1954). There are many low-lying rocky areas of Oronsay and its neighbouring islets that would have been suitable places for these clumsy, flightless birds to nest on.

Bones of the other species of auks (razorbills *Alca torda*, guillemots *Uria aalge* and puffins *Fratercula arctica*) are less numerous in the middens than those of the great auk, and it could be argued that in terms of relative importance in the diet (that is compared with non-avian sources of food) their contribution was negligible. However, this does not mean that their presence can be dismissed as fortuitous; perhaps they constituted a limited, seasonal resource. Again it seems reasonable to suggest that being colonial breeders they would have been caught when at their most concentrated, vulnerable and accessible, that is when breeding on land. There are no suitable cliffs for their colonies on Oronsay, nor on the southern part of Colonsay, nor would there have been any when the sea level was about 7 m higher. The only possible breeding cliffs are those where razorbills and guillemots (and many other sea birds) breed today, that is on the north-west coast of Colonsay below Beinn Bhreac and Bein Rhiabhach (see figure .2). Puffins no longer breed on Oronsay or Colonsay, but there is a green, turfy area high above the sea called Lamalum among the north-west cliffs of Colonsay that looks as though it would have been ideal for puffin burrows; there are no such areas on Oronsay or on the south coasts of Colonsay.

As the nearest place to Oronsay from which the smaller auks could have been easily obtained was probably the sea cliffs of north-western Colonsay, it seems likely that the other resources of the larger island would also have been exploited by the Mesolithic inhabitants of Oronsay: perhaps these included red deer.

4. GREY SEALS

Seal bones and teeth are found in all the middens from which mammal bones have been obtained (Cnoc Coig, Cnoc Sligeach, Caisteal-nan-Gillean I and the Priory midden). Although bones of the common seal (*Phoca vitulina*) and the grey seal (*Halichoerus grypus*) are mentioned from Cnoc Sligeach and Caisteal-nan-Gillean I (Lacaille 1954), all those identified from the new excavations, and from a re-examination of the surviving Cnoc Sligeach bones, are of the grey seal.

Around the western coasts of Scotland grey seals come to land (haul out) in order to moult in winter or in early spring using low-lying rocks and skerries, but it is only in the autumn, when they breed on various isolated islands, that they are land-based in any large quantity and are relatively easy to approach on foot. In the summer they are more or less dispersed at sea (Hewer 1974). As there are seal breeding colonies on the islets near Oronsay today the presence of their bones in the middens may not seem surprising, but at the time of the 7 m sea level Eilean nan Ron would have been under water and Eilean Ghaoideamal would have been very small. Were seals breeding on or near Oronsay, and if so where? What relevance does this have to the geography and history of the seal breeding grounds? The degree of attractiveness of seals to Mesolithic man and the influence of the likely situation of the breeding grounds to the distribution of the middens will also be discussed.

4.1 Location of the seal breeding colony in the Atlantic period

Future work on the seal bones from the midden will, it is hoped, shed light on their sex and age at death and help to show whether seals were caught during the breeding season or when hauled out at other times of the year. Seals can also be caught in nets at sea, but grey seals are so easy to obtain on land that this seems unnecessary to postulate in a Mesolithic context. Among the adult seal bones in the Cnoc Coig material there is an astragulus that is undoubtedly foetal and there are other bones that seem to be from very young animals, which allows one to suggest that both young and adult animals were being obtained from a breeding site.

The modern pattern of breeding behaviour of the seals of Eilean nan Ron and to a lesser extent of Eilean Ghaoideamal has been described in some detail by Hewer and Backhouse (1960) and Hewer's (1974) conclusion was that their behaviour on these islets represents a special "Inner Hebridean" type, in which, although the females use the tiny beaches at the level of high water springs for pupping, the bulls' territories are the erosional platforms that surround the islets and lie at about the present mean tide level, that is they are underwater at high tide. However on Oronsay these erosional platforms only exist at that particular level, and would have been submerged at a time of higher sea levels, so any breeding territories at that time would have had to have been organized differently. A second, or 'Outer Hebridean', type of territoriality which Hewer describes is one in which pupping and male territories are well above the high tide level on island beaches and on low-lying grassy or peaty hollows that are easily accessible from the sea. There is a third type where even the highest parts of quite large, though isolated, islands are used, such as North Rona and some of the Outer Farnes, but this may well be a relatively recent response to the rapidly increasing seal populations in British waters. If there were breeding colonies of seals on Oronsay in the Mesolithic it seems as though they would have had to be organized differently from today, probably in the "Outer Hebridean" fashion, as exemplified on the island of Shillay (Hewer 1957). It also seems that the use of the erosional platforms near Oronsay by seals is a behavioural response to lowered sea levels during the last 5,000 years, and of course to more intensive use of the island by man.

Which parts of Oronsay could have been used by breeding seals? One can immediately exclude any wide areas of gently sloping sandy beach and very steeply shelving rocky coasts, and Dubh Eilean can be excluded as access to virtually every dip and gully around its coasts is blocked by steep, raised, storm beaches. The pebble storm bars that make up the southern tip of Oronsay would not have been suitable. In the Outer Hebrides the sheltered eastern coasts are preferred by seals; possible places on Oronsay (see figure .1) include the rocky eastern coasts just south of Cnoc Sligeach (where narrow ancient beaches slppe steeply up to grassy areas below Leaba Mhor), the low area around Cnoc Coig, and the area to the south of Caisteal-nan-Gillean I, that is the south-eastern shore below Drium Mor. This seems to be the most likely spot if one takes Shillay as a model. A fairly steep erstwhile beach slopes up to and between flat rocks that were probably barer then than now, and there is extensive open grass land behind, complete with a possibie peat wallow in and around the lochans. It is possible that the whole of the southern part of Oronsay was a separate island in the Mesolithic, perhaps connected to Oronsay at low tide by a sandy strand (Jardine 1977), but this would not preclude the selection of this area as the most likely breeding place for grey seals at that time.

Intensive human occupation of a breeding island would soon frighten the seals away and it may be significant that three of the middens are on the main part of Oronsay. The middens of Caisteal-nan-Gillean I and II are however on the southern part. Seasonal evidence has been obtained for Caisteal-nan-Gillean II and suggests that it was used for fishing in the summer (Mellars 1978 & Mellars & Wilkinson forthcoming) and possibly for birding in the spring (auks); it may well be that people stopped fishing from the breeding island (or area) in the autumn and used it only for occasional seal hunting trips.

It is of course both possible and likely that seals hauled out on and around the Oronsay coast in the past as they do today (and the skerries that would have existed on the west of the island, particularly at Eilean an Uan, would have been suitable for this), so seals could have beem obtained when hauled out in early spring. However Mesolithic man was surely an opportunist (Mellars 1976) and wherever possible would have used animal resources in places and at times of the year when they were at their most concentrated and when they could be obtained with a minimal investment of time and effort, that is, when they were in their breeding territories.

4.2 Why seals?

Laughlin (1977) has produced a list of the many uses to which seals are put by Eskimos (reproduced in table 2). I would like to stress the efficacy of seal skin in the construction of boats and as clothing. It is salutary to remember that it was the substitution of India rubber for seal skin boots, and of paraffin for seal-oils, that helped to put an end to the nineteenth century persecution of the grey seal (Darling 1940). Seals also provide a lot of food, the pups are quite small, but the average weight of adult bulls is 225 kg ($4\frac{1}{2}$ cwt) and their weight can exceed 6 cwt (about 300 kilos) (Prime, p.c.).

Multiple Use of Resources

(Eumetopia jubata Northern or Steller sea lion)

Part of Animal	Partial List of Uses
1. Hide	Cover for kayak and umiak; line for harpoon
2. Flesh	Food, for humans
3. Blubber	Food: eaten with meat, rendered for oil
4. Organs (heart, liver, spleen, kidney)	Food
5. Bones	Ribs for root diggers, humerus for club, baculum for flaker
6. Teeth	Decorative pendants
7. Whiskers	Decoration of wood hunting hats and visors
8. Sinew	Back sinews used for sewing, lashing, cordage (less desirable than sinew of whale or caribou)
9. Flippers	Soles used for boot soles; contents gelatinized in flipper and eaten
10. Pericardium	Water bottle, general-purpose container
11. Esophagus	Parka, pants, leggings of boots, pouches
12. Stomach	Storage container (especially for dried salmon)
13. Intestines	Parka, pants, pouches

TABLE 2 The uses made of parts of the Northern Sea Lion (Eumetopia jubata) by Eskimos. (From Laughlin 1977).

5. THE ATTRACTIONS OF THE LITTORAL ZONE AND OF ORONSAY IN PARTICULAR

Alan Osbourne (1977) has attacked the idea that the littoral zone in general was an attractive environment for early man, arguing that as a resource base it is second-rate, being both labour-intensive and low-yielding, to be utilized only in times of seasonal scarcity or famine or when the human carrying capacity of the adjacent land had been exceeded. He based these arguments largely on the relative quantities of protein produced by terrestrial herbivores compared to marine invertebrates: for example in the New World 83,422 mussles yield the same quantity of protein as <u>one</u> white-tailed deer. He goes on to show that in terms of protein content marine mammals are less productive than terrestrial herbivores and are less efficient sources of energy, as, being carnivores, they are higher in the food chain. I suggest that it depends entirely on local conditions. Osbourne points out that the seals of the Peruvian coast (that is, the sea lions) are extremely difficult to hunt, and this contrasts with the ease with which grey seals can be caught at their breeding stations. I would like to present the hypothesis that, unlike some other marine mammals, grey seals in the breeding season were a "low-investment, high-return" resource of the greatest importance, and that the high calorific and nutritional value of their <u>fat</u> (not to mention its many non-dietary day-to-day uses) was the seals'chief allurement for people living in cool, wet environments. Indeed it is the use of parts of the coasts of Oronsay and Colonsay by terrestrial breeding colonies of otherwise diffused pelagic popolations of birds and mammals that must have formed the islands' main attraction for Mesolithic hunters.

Islands and coastlines of complicated topography have the added advantage of a long shore line relative to the area of land within a given territory (Mellars 1978), and the variegated, gently sloping, rocky shores of Oronsay and the south-west of Colonsay with their huge limpet and crustacean populations would have provided a dependable, if not very nutritious source of food, particularly at times of seasonal or occasional scarcity or when rough seas prevented fishing, birding or sealing.

If one adds the ecotonal nature of the Scottish coast line with its opportunities for the exploitation of rocky and sandy shores, inshore seas and cliff nesting sites, as well as of the terrestrial plants and animals of the hinterland, or in this case, of the interior of the islands, one is surely approaching a situation that was in many ways ideal for hunters and gatherers. And of course there were the seals: in no way could the resources and environment of Oronsay be described as second-rate.

6. SUMMARY

The main alternative hypotheses regarding the presence of red deer bones in the Oronsay middens are (1) that they came from animals living on Colonsay, and (2) that they came from animals living on Islay or Jura. In the course of discussion of these hypohteses it became clear that Colonsay should be included when assessing the likely resource base of the Mesolithic inhabitants of Oronsay, and that both the present underwater topography around the islands and their very limited terrestrial fauna suggest that Colonsay and Oronsay have been isolated since before the beginning of the Holocene. It is hoped that bone element analysis and a comparison of the size of the red deer bones from both the old and the new excavations with those from contemporary sites on Jura and the mainland will allow these hypotheses to be tested, and will also help to establish whether the deer were used primarily as a source of bone and antler for tools or whether they were eaten as well.

It is suggested that as grey seals were probably breeding on Oronsay itself they were hunted there in the breeding season; this hypothesis will be tested by studying the age distribution of the seal bones in the Cnog Coig midden.

7. ACKNOWLEDGEMENTS

The help and criticism of earlier drafts of this paper by Paul Mellars, Graham Jardine, D. B. Gilbertson, John Prime and K. M. Backhouse, are gratefully acknowledged; the responsibility for the views expressed is my own. I thank Don Bramwell for access to unpublished information on bird bones. This essay is dedicated to Joe Banks, who walked the islands with me.

BIBLIOGRAPHY

Anderson, J., 1898. Notes on the contents of a small cave or rock shelter at Driumvargie, Oban; and of three shell mounds in Oronsay. Proc. Soc. Antiq. Scot. 32: 298-313.

Arnold, H. R., (Ed.), 1973. Provisional Atlas of the Amphibians and Reptiles of the British Isles. Biological Records Centre, Abbots Ripton.

Arnold, H. R., 1978. Provisional Atlas of the Mammals of the British Isles. National Environment Research Council, Abbots Ripton.

Binford, L. R. and Bertram, J. B., 1977. Bone frequencies - and attritional processes. In Binford, L. R. (ed.), For Theory Building in Archaeology pp. 78-153. Academic Press, London.

Birks, H. J. B., 1973. Past and Present Vegetaion of the Island of Skye. University Press, Cambridge.

Birks, H. J. B., 1977. The flandrian forest history of Scotland: a preliminary synthesis, In: Shotton, F. W. (ed.), British Quaternary Studies-Recent Advances, Pp. 119-135. University Press, Oxford.

Bishop, A. H., 1914. An Oronsay shell-mound: a Scottish pre-Neolithic site. Proc. Soc. Antiq. Scot. 48: 52-108.

Clutton-Brock, J., (in press). A Natural History of Domesticated Animals.

Corbet, G. B., 1961. Origin of the British insular races of small mammals and of the Lusitanian fauna. Nature 191: 1037-1040.

Cunningham Craig, E. H., Wright, W. B. & Bailey, E. B., 1911. The geology of Colonsay and Oronsay, with part of the Ross of Mull. Mem. Geol. Surv. Scotland 35: 1-74.

Darling, F. F., 1940. Island Years. Bell, London.

Evans, J. G., 1971. Habitat change on the calcareous soils of Britain: the impact of Neolithic man, In: Simpson, D. D. A. (ed.), Economy and

Settlement in Neolithic and Early Bronze Age Britain and Europe. Pp. 27-73. University Press, Leicester.

Fisher, J. and Lockley, R. M., 1954. Sea Birds. Collins, London.

Grieve, S., 1885. The Great Auk or Garefowl. (Alca impennis, Linn.). Jack, London.

Grigson, C., 1969. The uses and limitations of differences in absolute size in the distinction between the bones of aurochs (Bos primigenius) and domestic cattle (Bos taurus). In Ucko, P. J. and Dimbleby, G. W. (eds.), The Domestication and Exploitation of plants and animals. Pp. 277-294. Duckworth, London.

Grigson, C., (in press). The Mesolithic fauna. In Simmons, I. G. & Tooley, M. J., British Prehistory: the Environment. Pp. 110-124. Duckworth, London.

Hewer, H. R., 1957. A Hebridean colony of the grey seal (Halichoerus grypus, Fab.) Proc. Zool. Soc. Lond. 128: 23-66.

Hewer, H. R., 1974. British Seals. Collins, London.

Hewer, H. R. & Backhouse, K. M., 1960. A preliminary account of a colony of grey seals, Halichoerus grypus, Fab. in the southern Inner Hebrides. Proc. Zool. Soc. Lond. 134: 157-195.

Jardine, W. G., 1977. Location and age of Mesolithic coastal occupation sites on Oronsay, Inner Hebrides. Nature 267: 138-140.

Jardine, W. G., 1978. Radiocarbon ages of raised-beach shells from Oronsay, Inner Hebrides, Scotland: a lesson in interpretation and deduction. Boreas 7: 183-196.

Lacaille, A. D., 1954. The Stone Age in Scotland. Oxford University Press, London.

Laughlin, W. S., 1977. Hunting: an integrating biobehaviour system and its evolutionary importance. In Lee, R. B. and de Vore, I. (eds.), Man the Hunter. Pp. 304-320. Aldine, Chicago.

Loder, J. de V., 1935. Colonsay and Oronsay in the Isles of Argyll, their history, flora, fauna and topography. Oliver, & Boyd, Edinburgh.

McNeill, M,, 1910. Colonsay, its Plants, Climate, Geology, etc. Douglas, Edinburgh.

Mellars, P., 1976. Settlement patterns and industrial variability in the British Mesolithic. In Sieveking, G. de G., Longworth, I. H. & Wilson, K. E. (eds.), Problems in Economic and Social Archaeology. Pp. 375-399. Duckworth, London.

Mellars, P., 1978. Excavation and economic analysis of Mesolithic shell middens on the island of Oronsay (Inner Hebrides). In Mellars, P. (ed.), The Early Postglacial Settlement of Northern Europe. Pp. 371-396. Duckworth, London.

Mellars, P. & Wilkinson, M. R., (forthcoming). Fish otoliths as an indication of seasonal occupation in Mesolithic shell middens.

Mitchell, B., Staines, B. W. & Welch, D., 1977. Ecology of Red Deer. Institute of Terrestrial Ecology, Cambridge.

Osbourne, A. J., 1977. Standloopers, mermaids and other fairy tales: ecological determinants of marine resource utilization—the Peruvian case. In Binford, L. R. (ed.), For Theory Building in Archaeology. Pp. 157-205. Academic Press, London.

Paul, C. R. C., 1976. The non-marine mollusca of Colonsay and Oronsay. J. Conchology 29: 107-110.

Peacock, W. R. B., 1978. Probablistic sampling in shell middens: a case study from Oronsay, Inner Hebrides. In Cherry, J., Gamble, C. & Shennan, S. (eds.), Sampling in Contemporary British Archaeology. Pp. 177-190. B.A.R. 50, Oxford.

Simmons, I. G., 1975. The ecological setting of man in the Highland Zone, In Evans, J. G., Limbrey, S. & Cleere, H. (eds.), The effect of man on the landscape: the Highland Zone. Pp. 57-63. C.B.A. Research Report no. 11, London.

Spenser, P. J., 1975. Habitat change in coastal sand-dune areas: the molluscan evidence, in: Evans, J. G., Limbrey, S. & Cleere, H. (eds.), The effect of man on the landscape: the Highland Zone. Pp. 96-103. C.B.A. Research Report no. 11, London.

Sutcliffe, A. J., & Kowalski, K., 1976. Pleistocene rodents of the British Isles. Bull. Brit. Mus. Nat. Hist. (Geol.) 27, no. 2.

Uerpmann, H.-P., 1973. Animal bone finds and economic archaeology: a critical study of 'osteo-archaeological' method. World Arch. 4: 307-322.

THE USE OF FISH IN RESOURCE SCHEDULING AND SEASONALITY STUDIES IN TEMPERATE LATITUDES

By M. Wilkinson

> "faunal research certainly should be directed along a number of paths of inquiry in determining the seasonality of exploitation" (Smith 1976, 290).

The seasonal variation of climate in temperate regions exerts a major influence on the available food supply for the human, and animal, populations that live there. In response they have developed a range of behavioural adaptations to deal with this situation, the most widespread of which include: (i) the restriction of population size to a level that can be supported during the period of least availability; (ii) intensification of the exploitation system; (iii) broadening of the subsistence base to include additional resources; (iv) exploitation of a larger territory; (v) or the development of food storage techniques.

Almost all of these responses entail some seasonal variations in the exploitation of resources, and archaeologists have shown increasing interest in understanding such patterning. Although this interest has been most apparent in studies of groups with food-collecting economies (hunting, gathering and fishing), it is also fundamental to food-producing systems. The labour requirements of crop-production are highly seasonal and they have a very restricted period of yield, while the same can apply to stock. In many instances food storage helps to overcome these fluctuations in yield for farming groups, while the low level of its adoption in most food-collecting economies results in much greater seasonal patterning. Variation in the productivity of potential foods through the year is responded to by specialisation on resources, or by diversification to include a range of foods, both of which usually entail the occupation of more than one site during the course of the year.

> "The gatherer-hunter-fisher subsistence system is usually a seasonal circuit or alternation of successive sites and their catchment areas, which exploit varied combinations of productive ecological sectors" (Clarke, 1976, 469).

With this in mind no study of the economy of food-collecting populations can hope to understand the patterns presented without considering seasonal variations in the subsistence strategy.

MODELLING SUBSISTENCE STRATEGIES

Before considering the seasonal patterning of fish exploitation, it is useful to outline the principles used in the modelling of subsistence systems, which apply both to the seasonal use of particular resources and to the overall proportions in which they are exploited. The aim of the method is:

"to evaluate the most favourable periods for exploiting the different species from a biological as well as a hunting-strategic point of view" (Mikkelsen, 1978, 89).

For this a number of behavioural features of the resource are considered, together with certain cultural elements of the exploitative population, to establish the most efficient system of exploitation. Most studies appear to agree on the criteria they consider fundamental in such decision-making:

i. Mikkelsen (1978) considered changes in animal densities, behaviour patterns and the availability of animals to people.

ii. Poiner (1976), in a comparative survey of coastal hunter-gatherers, identified the range of resources available, their abundance and their predictability; accessibility is also an important factor but is related to the other variables and to the available technology.

iii. Bulmer's (1976) study of the Kalam of New Guinea found that the extent to which resources were utilised depended upon their population density; the extent and proximity of their habitats to the homestead; the degree of domestication; their habits, in terms of whether they were easy or difficult to capture; the skills, knowledge and technology available; the seasons in which they were hunted; and the economic and ritual values attached to them.

iv. Jochim (1976) recognises two major goals to be achieved in resource use scheduling: the reliability of resources and the limiting of effort. Each of these aims can be described in terms of a number of attributes he assigns to each resource: weight, density, aggregation size, mobility and non- food yield.

$$\text{Reliability} = \frac{\text{weight x non-food yield x density}}{\text{mobility}}$$

$$\text{Limiting of effort} = \frac{\text{weight x non-food yield x aggregation size}}{\text{mobility}}$$

In addition he proposes several secondary goals, such as good-tasting foods, foods with a high fat content, variety, prestige and maintenance of sex roles, and also includes 'cultural' factors like food storage, imposed limitations on procurement and aids to procurement.

The ranking used by Jochim illustrates one of the difficulties in such modelling. They are all based on optimisation and energy efficiency, and so tend to be biased toward the production side of the model. Resource use scheduling is based on decision-making and often includes a series of possible options favouring alternative goals. For instance, stability and reliability may be of particular importance in certain environments and will favour a 'generalised' economy as opposed to a more efficient 'specialised' regime of optimal exploitation. This can be illustrated from the combination of fishing and farming in crofting economies (Coull,1971). In Norway the two activities have been easily integrated, with most fishing activity in winter when the demands of farming are least, whereas in Scotland the summer fishing places a heavy burden on the population with much of the farmwork falling on the women (Grant, 1961, 262); however, both of these situations may be reflecting commercial pressures rather than subsistence decisions (Coull,1971; Paine, 1957, 101). The problems inherent in trying to rank goals can be summarised in a further quotation:

> "duration of availability, season of availability and congruency of fish with other resources would determine the degree to which anadromous fish would be exploited by humans, regardless of abundance" (Schalk, 1977, 242).

THE ANNUAL CYCLE OF FISH EXPLOITATION

The most important factors in determining the intensity with which a resource is exploited appear to be its abundance, predictability and accessibility, and its relationship to other resources. Before these are considered some basic considerations are of relevance. Firstly, fishing must be regarded essentially as a 'hunting' activity where control over the environment and the resource is strictly limited (Andersen & Wadel, 1972, 153). Secondly, it may be misleading to treat fish as a single resource, as the behaviour patterns of species do vary considerably. The final point, of direct importance to the annual cycle, is that fish are cold-blooded and so their metabolism etc. is closely related to the temperature of the surrounding water. Hence the annual climatic cycle in temperate regions produces a marked seasonal patterning in the life of fishes.

Turning to the four criteria listed above, very little can be made of the relationship to other resources as this depends as much on external factors as on the intrinsic qualities of the resource; in terms of the annual pattern the most important point is whether resources can be used consecutively, or are coincident and therefore require priority decisions.

Accessibility: This will depend upon the available technology and the species being sought, but generally the best locations to minimise effort and risk are those closest to the shore or in shallow/confined stretches of water. Regardless of the availability of the fish there is also the influence of such factors as weather on the fishing effort (Landberg, 1975, 159).

Predictability: The two major components of this are the overall stability of the system and the regularity of timing within the cycle. Aquatic environments are generally fairly stable, particularly the marine system (Odum, 1975, 171), and this applies also to their fish populations. The very marked regularity of events in the annual life of fishes has been stressed by several authors and must be closely related to their dependence on the temperature of the surrounding water (Cushing,1969; Schalk, 1977).

Abundance: There are two main situations that result in periods of maximum abundance. The first applies only to open water systems and is the time when the largest number of individuals are to be found in an area, including any migratory populations. The other is a smaller scale repetition of the first and is based on the congregation of individuals within the area. In both the result is to minimise the area that has to be searched to locate the resource.

The most important point about all of these attributes is their marked seasonal variability. Any intensive exploitation of the resource is likely to be concentrated on the times when the fish are present in greatest quantity, are most accessible and in good condition. In terms of fish behaviour these situations are respectively found when:

i. fish aggregate either to take advantage of favourable conditions (as on spawning and feeding grounds), or when they undertake migrations and so pass through a body of water in large numbers.

ii. they enter inshore or shallow waters; these offer protection and rich food resources, and are therefore very attractive to fish, as well as being the most easily accessible to fishermen.

iii. they are in prime condition; fish undergo considerable changes in weight etc., with peaks in 'condition' after feeding, and prior to major events such as spawning and migrations.

In many instances some or all of these conditions are coincident, indicating clear optimal periods for intensive exploitation. No attempt will be made here to offer a detailed account for any one species or area, but in general terms certain circumstances can be identified:

Spawning: One of the most important points about this event is the predictability of its timing (Cushing, 1969); most species spawn in the spring or early summer, apart from some of the salmonids who lay their eggs in the autumn. For many species it is associated with aggregation on spawning grounds and may be preceded or followed by migration. In a large number of cases these grounds are in shallow waters; the most extreme examples of this are anadromous fishes like the salmon that live in the sea but enter freshwater to spawn. As the time for spawning approaches most fish are in good condition, with a high fat content as well as the roe which is a desirable product from many species.

Feeding: Behaviour during the main summer feeding period is more variable, but many species aggregate either in shoals in open water or on feeding grounds; the condition and weight of the fish increases through the feeding season.

These favoured situations are counterbalanced by periods of low availability. For many fish this is the winter when low temperatures inhibit activity and limit the food supply; some species become inactive at this time while others disperse into deeper waters. Winter is also the season when inclement weather is likely to upset fishing activity to the greatest extent.

Two final points ought to be made to broaden the scope of this discussion. It has been based on the patterns of behaviour of individual species although most fish faunas are mixed. The second point is that the discussion has concentrated on an 'intensive' fishery, where it is possible to assume that optimal efficiency is important; on a different level, where fish is used to supplement the diet, provide variety or make up a shortfall in regular foods, such criteria need not be applicable.

ARCHAEOLOGICAL APPLICATION OF RESOURCE SCHEDULING

The listing of species and their inferred season of exploitation is too familiar in the archaeological literature to require illustration, but it is noteworthy that fish are one of the prime sources for such interpretations. The method usually relies upon the most obvious time of capture of individual species, and sometimes on the absence of particular fish, but can be extended to examine associations of species or the behaviour of specific age groups. All of these studies assume that exploitation occurred at an 'optimal' period and make use of modern behaviour patterns, although neither of these assumptions is necessarily valid. The possibility of fishing at other times of the year is sometimes mentioned (Kaelas, 1976, 134) but deserves greater consideration. Some indication of whether alternative strategies were in operation may be gained from the nature of the fish remains and the other resources exploited. The quantity of fish, together with the diversity of species can indicate the intensity of the fishing activity, and the seasonality of the other foods may point to the time of fishing.

It is necessary to stress the point that, with certain exceptions, most species have a long season during which they can be caught (absolute availability) with a much shorter period for optimal exploitation (relative availability); therefore interpretations using the latter are only statements of probability. This raises two points neither of which has been fully investigated; firstly there is a need to consider exploitation strategies not based on optimisation, and secondly there is the possibility of quantifying these probabilities. By calculating the proportion of the annual cath taken in each month the likelihood of capture at any time of the year can be estimated; this could be based upon known fishery statistics (Leach, 1979) or on ecological modelling (Jochim, 1976).

The general pattern of fish exploitation presented by Jochim agrees with the conclusions reached above, but there are a number of points that deserve consideration. The most important of these is the inclusion of all fish within a single resource group, when species with different behaviour patterns, sizes etc. are obviously represented. In this context it is worth noting the underestimation of the autumn-spawning fishes in the modelling of the Round Lake Ojibwa system (44). Following from this it is possible to question the 'average' values that are used for the resource attributes. This is not to question the value of resource use scheduling but to suggest that it be applied in rather greater detail in future modelling.

At the same time it is clear that there is a need to develop alternative methods for determining seasonality. The various lines of such enquiry are summarised by Smith (1976, 288):

i. niche reconstruction of animal species.

ii. research into ethnohistorical records.

iii. direct archaeological evidence for seasonality.

The first of these covers the modelling of resources considered above and the second needs little explanation. The final source has received least attention and only for shellfish has its potential been appreciated. In the same way these techniques can be applied to other resources and fish are one of the most promising fields, given the work that has been done already in fisheries science.

PRINCIPLES OF AGE DETERMINATION IN FISHES

In both biological and fisheries studies great importance has been attached to ageing, in order to understand the behaviour and population structure of fish stocks. Apart from some controlled experiments, with marked or captive fish, the vast majority of age determinations have to be based upon intrinsic properties of the fish. Two main techniques exist, both based upon the manner of fish growth, and as this differs in fundamental ways from that of other vertebrates, it is necessary to consider this first.

Growth in Fishes: There are three important aspects of growth to consider: the pattern through life; the variability of growth; and the pattern of growth within a year. In generalised terms fish grow throughout their lives, following a sigmoid growth curve, whereby annual increments increase until maturity after which the growth rate declines. However, in terms of relative growth, that is the size of an increment as a proportion of the original size, there is a continuous fall through life; for instance cod fry increase in size by 3-4 times during their first summer while yearlings only double their lengths in the same period (Wollebaek, 1900, 104). Several periods can be recognised in the life of fishes based on the character of growth (Nikolsky, 1963, 201-2), with the clearest division lying between the period prior to maturity (rapid linear growth) and that following it (declining increments).

Growth occurs in response to a series of internal physiological factors and extrinsic physical ones such as temperature, food, photoperiod, salinity etc. The most important external influences are the temperature and the availability of food. Fish can respond to variations in these conditions by adopting different rates of growth to the extent that there are no prescribed size ranges for age groups or the ultimate size of the fish (indeterminate growth) (Iles, 1974, 331). This is most apparent in closed freshwater systems where 'stunted' populations are not uncommon. This variability in growth rates is apparent not only between populations but also within them, especially under stress conditions (Nikolsky, 1963, 206).

With the close relationship that exists between fish growth and environmental conditions, growth is bound to follow a seasonal pattern in temperate latitudes. This is not easy to delimit but appears to fit a sigmoid curve thus transforming the overall growth curve into a series of waves of growth.

> "It appears to be a common feature of the seasonal growth curves that they are sigmoid in shape. There is a period during which growth accelerates, then one of almost regular rapid growth, followed by one during which growth declines" (Iles, 1974, 337).

The year can be divided into two main seasons, with active growth during the spring and summer, followed by minimal increases over the winter; as an example, cyprinids grow between June and September, with none recorded from November to March (Steinmetz, 1974, 148). The most significant point in the year occurs in the spring when growth resumes after the winter break; the fall-off at the close of the active growing season is more gradual.

Age Determination in Fishes: Before describing the two main ageing techniques, it will be useful to briefly explain some of the terms used. There are several systems for naming age groups, but the basic ones are the year class-all fish born in a particular year (1978 class etc.)—and the age groups —which covers all fish of the same age (yearlings, second year fish etc.). The most straightforward designation of age groups is to refer to fish as first year, second year...or to use the corresponding fisheries system of O-group, I-group... All of these procedures are simplified by the restricted spawning season of fishes which produces clear age groupings; a further simplification is the adoption of January 1 as the official birthday of fishes.

The easiest method of ageing is the Petersen length-frequency technique (Chugunova, 1963, 110-112), which is based on all individuals of the same age being approximately the same size. Given the restricted spawning season and similar growth rates for all individuals, then each age group should appear as a mode when plotted on a length-frequency histogram. This is a quick and simple procedure and is much used in fisheries work but is rather crude and has certain limitations. For instance growth must be uniform and so rules out most mixed populations; equally important is the declining growth rate of older fishes, which leads to progressive overlapping of length frequencies (Weatherley & Rodgers, 1978, 56). Therefore it is most suitable for use on younger age groupings.

The second method is based upon the examination of growth rings in the 'hard parts' (i.e. bones, including scales and otoliths) of fishes. Bones must increase in size relative to the growth of the fish and so by analogy there should be two main patterns of bone growth: a series of annual increments corresponding to the sigmoid growth curve through life; and an annual pattern with most bone growth during the spring-summer season. The new bone is added to the exterior surfaces, respecting their existing shape, and in simplified terms is almost two-dimensional on flat, plate-like structures (scales, cleithra, opercula etc.) and concentric rings on cylindrical forms (vertebrae, spines etc.). The deposition of material within the year is normally concentrated in the spring-summer period; in the pike 80% of linear growth of the cleithrum occurs in 22% of the year (Casselman, 1974, 13). These variations in the rate of deposition are associated with changes in the composition of the material laid down:

> "The periods of slow growth are imprinted on the skeleton in the form of rings or stripes, which are pale in reflected light. Conversely, the periods of faster growth are characterised on the scales and skeletons by wide fields or rings, which are dark in reflected light..." (Nikolsky, 1963, 190).

This distinction of two zones within the annual increment is most clearly seen in the otoliths, but exists in other bones also; it is the result of differences in the structure of the inorganic material together with variations in the amount of the organic component. However, the most obvious distinction is not within the annual increment as the fast zone grades into the slow one, but between the years, when rapid growth in spring follows the winter lull. These breaks are counted to give the number of annual increments and therefore the age of the fish. The technique is widely used in fisheries studies and its validity has been established many times by comparison with independent age determinations (Blacker, 1974b, 87-90). However, it is important to introduce a note of caution at this stage as the formation of annual rings is an internal physiological process and has not yet been fully understood (Bagenal, 1974, v). Therefore, it is not safe to assume that the technique will work on any group of material until this has been tested (Blacker, 1974a, 113).

When the method has been proven there are a number of practical problems in age reading to consider; these can be summarised in three categories:

i. False or secondary rings: When the growth pattern of the fish is disturbed false 'winter' rings may be formed; this is a common occurrence and has various causes e.g. spawning (de Bont 1967, 71). These can be confused with the true annuli but are usually less well formed.

ii. Missing rings: This can occur if growth material is subsequently removed, as in the resorption of scales, or if growth is so slow that there is no noticeable distinction between summer and winter rates.

iii. Annulus reading is a subjective procedure and leads to varying interpretations, although this can be minimised (Blacker, 1974a). Rings can be very difficult to read especially if they are obscured by subsequent growth or by the crowding of rings in later life.

These problems demonstrate the need for care in the use of the technique (de Bont, 1967, 76). One of the most important points is the choice of the most appropriate bone as this varies between species. The elements that are most extensively used are scales, for the Salmonidae and Cyprinidae (Chugunova, 1963, 17-43), and otoliths of the Gadidae and Pleuronectidae (Williams & Bedford, 1974). Other bones are also used including the operculum, cleithrum, vertebrae and fin spines.

Archaeological Applications: These techniques can be used to investigate the age composition of the catch and the season of death of the fish, although only a few examples of such studies can be cited.

i. The age composition of the catch can be determined from length frequencies of the reconstructed sizes of the fish or by counting the annuli on bones. However, this is of only restricted use, possibly allowing inferences about age-related behaviour patterns and enabling the reconstruction of the age and size structure of unexploited fish stocks

ii. Of much greater potential is the possibility of determining the season of death of the fish. This can be done with fish size if growth is sufficiently fast and rates of increase are known but growth rings are more widely applicable. By assessing the amount of growth since the most recent annulus it is possible, once the pattern of bone deposition is known, to estimate the time of year that death occurred. Where two recognisable rings are laid down within the year resolution is enhanced, as in otoliths.

An example of such a study is that by Casteel (1972) in which four caches of fish-bone from a midden were tested for seasonality; the amount of growth beyond the annulus on a series of vertebrae was described in terms of three seasons and each cache showed a clear predominance of one time of year.

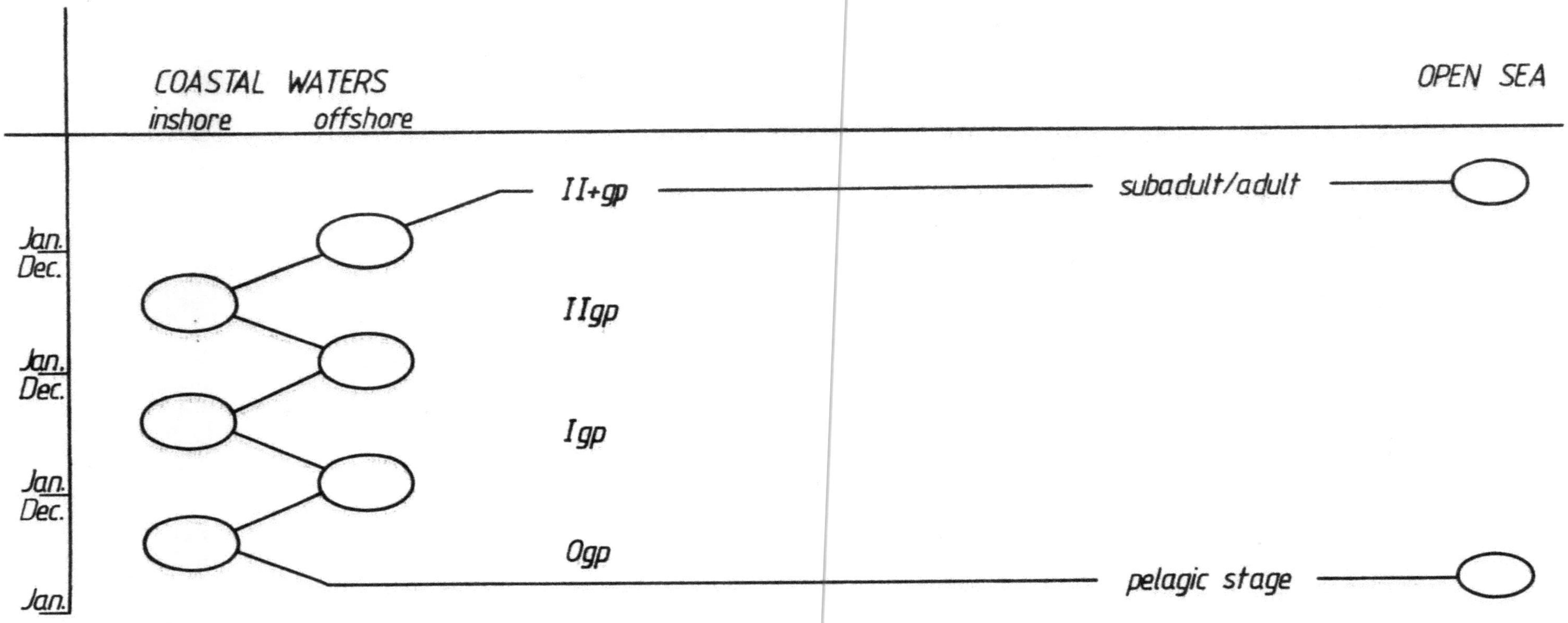

Figure 1. Seasonal movements of *Pollachius virens* L. in inshore waters.

But in each case some 12% to 14% of the readings showed a different pattern (Casteel, 1972, 409) which was interpreted as erroneous readings of contamination of the samples. However, two other explanations can be suggested which point to the limitations of the method. If different species or age groups within a species are included then as the annuli may be formed at different times of the year, all the bones will not show the same stage of development. Secondly, annulus formation is not a single simultaneous event throughout a population, but varies vetween individuals and so some level of variability is to be expected (Trout, 1954, 100; Gambell & Merstorff, 1964, 395). As pointed out by Wheeler (1978) care is needed is using this technique, but as Casteel demonstrated it has very great potential.

RESOURCE SCHEDULING AND THE ORONSAY MIDDENS

The discussion above provides the framework for a major element in the analysis of the fish remains from a series of Mesolithic shell middens on the island of Oronsay, (Inner Hebrides). Seasonality, both in the pattern of resource exploitation and in the sites occupied is taken to be basic to the groups that left these traces. As has been demonstrated fish provide one of the clearest indicators of such behaviour patterns and can be used in various ways. Here the application of resource scheduling is considered and its limitations in seasonality studies is discussed.

The shell middens all date to the fourth millennium b.c. and are being investigated in a project directed by Dr. P. Mellars (Mellars, 1978). There are at least five major shell middens on the island and this raises problems of their interrelationships, in addition to the considerations of the wider site system. They contain evidence for the exploitation of a wide range of economic resources including large quantities of fish. Over a dozen species are present but the material is dominated by one species, the saithe or coalfish, <u>Pollachius virens</u> (Reinsch, 1976; Wheeler, 1969. 168-9), which is present in such numbers as to imply a fishery at a time of abundance.

There are three main life stages in this species, of which only the littoral phase is represented in the sites. This covers fish in the first three years of life, between the pelagic fry stage and the return to deeper water with the approach of maturity. In this inshore period there are marked patterns of behaviour, the main spatial and temporal features of which are summarised in Figure 1. Simply in terms of 'absolute' availability there is a distinction between the winter quarter of the year when the fish move out into deeper water and so are less accessible, and the remainder of the year.

The 'relative' availability of the species within these nine months is however subject to considerable variations. If attributes such as the ratios of the different age groups, the abundance of individuals within an age group and the size of individuals, are considered in the model then a detailed pattern of changing availability throughout the year is built up (Figure 2). While numbers of individuals fall through the year the avarage size of individuals is rising and there is a constantly changing ratio of the different age groups. An exploitation system to maximise meat yield would only need to balance these variables in terms of total productivity to determine the optimal time for fishing.

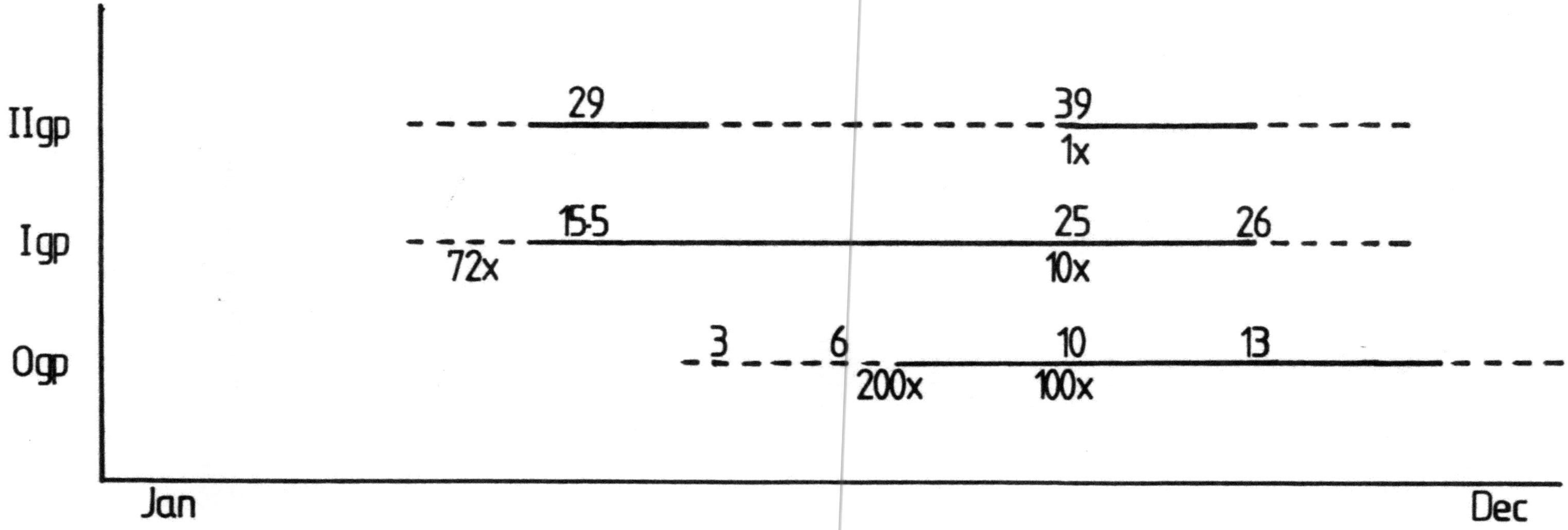

Fig. 2 Availability of Pollachius virens L. in inshore waters, based on the Faroes (after Bertelsen 1942).
29 – average size of individuals (cm)
72x – average number of individuals per haul

However, the changing structure of the resource through the year suggests that if other goals are important or other constraints exist, then the optimal period for exploitation may change. In this context it will be necessary to consider:

i. the method of fishing: fishing gears are size-selective so it may not be possible to fully exploit the resource at all times; also there is the 'efficiency' of the gear in terms of the amount of fish caught set against the effort of construction and fishing.

ii. the way the catch is utilised: the size and numbers of fish will determine or influence the ways in which they are processed, the time involved, and the uses to which they are put.

If the exploitation of this resource was attempting to maximise either the number of fish or the size of individual fish then there would clearly be different optimal periods. On this basis resource scheduling has proved inadequate in indicating the season of activity, as there is a possible nine month season, and so it has been essential to develop alternative techniques, based on fish size (Mellars & Wilkinson, forthcoming) and otolith growth rings. On the other hand the method has given an indication of the structure of the resource and raised important questions about the aims and methods of this particular fishery, and the possible alternative strategies on which it could have been based.

BIBLIOGRAPHY

Andersen, R. & Wadel, C., 1972. Comparative problems in fishing adaptations. In R. Andersen & C. Wadel (eds.), North Atlantic fishermen anthropological essays on modern fishing Newfoundland Soc. Econ. Pap. 5, 141-165.

Bagenal, T. B., 1974. Introduction. In T. B. Bagenal (ed.), The Ageing of Fish. Unwin, Old Woking.

Bagenal, T. B., (ed.) 1974. The Ageing of Fish Unwin, Old Woking.

Bertelsen, E., 1942. Contributions to the biology of the Coalfish (Gadus virens L) in Faroe waters. Medd. Komm. Dan. Fisk. Havunders. (Ser. Fisk.) 11 (2), 1-68.

Blacker, R. W., 1974a. The ICNAF cod otolith photograph exchange programme. In T. B. Bagenal (ed.), 108-113. The Ageing of Fish. Unwin, Old Woking.

Blacker, R. W., 1974b. Recent advances in otolith studies. In F. R. Harden Jones (ed.), Sea Fisheries Research, 67-90. Elek, London.

Bulmer, R., 1976. Selectivity in hunting and in disposal of animal bone by the Kalam of the New Guinea Highlands. In G. de G. Sieveking, I. H. Longworth, & K. E. Wilson (eds.), Problems in Economic and Social Archaeology, 169-186. Duckworth, London.

Casselman, J. M., 1974. Analysis of hard tissue of pike Esox lucius L with special reference to age and growth. In T. B. Bagenal (ed.), 13-22. The Ageing of Fish. Unwin, Old Woking.

Casteel, R. W., 1972. Some archaeological uses of fish remains Am. Ant. 37, 404-419.

Casteel, R. W., 1976. Fish Remains in Archaeology Academic Press, London.

Chugunova, N. I., 1963. Age and Growth Studies in Fish Israel Program for Scientific Translations, Jerusalem.

Clarke, D., 1976. Mesolithic Europe: the economic basis. In G. de G. Sieveking, I. H. Longworth & K. E. Wilson (eds.), Problems in Economic and Social Archaeology, 449-481, Duckworth, London.

Coull, J. R., 1971. Crofter-fishermen in Norway and Scotland. University Dept. of Geography, Aberdeen.

Cushing, D. H., 1969. The regularity of the spawning season of some fishes J. Cons. int. Explor. Mer 33, 81-92.

de Bont, A. F., 1967. Some aspects of age and growth of fish in temperate and tropical waters. In S. D. Gerking (ed.) The Biological Basis of Freshwater Fish Production, 67-88. Blackwell, Oxford.

Gambell, R. & Merstorff, J., 1964. Age determination in the Whiting (Merlangius merlangus L) by means of the otoliths J. Cons. int. Explor. Mer 28, 393-404.

Grant, I. F., 1961. Highland Folk Ways. Routledge & Kegan Paul, London.

Iles, T. D., 1974. The tactics and strategy of growth in fishes. In F. R. Harden Jones (ed.), Sea Fisheries Research, 331-345. Elek, London.

Jochim, M. A., 1976. Hunter-Gatherer Subsistence and Settlement. Academic Press, London.

Kaelas, L., 1976. Pitted Ware culture: the acculturation of a food-gathering group. In S. J. de Laet (ed.), Acculturation and Continuity in Atlantic Europe Diss. Archaeol. Gandensis 16, 130-141.

Landberg, L. C. W., 1975. Fishing effort in the aboriginal fisheries of the Santa Barbara region, California: an ethnohistorical appraisal. In R. W. Casteel & G. I. Quimby, (eds.), Maritime Adaptations of the Pacific, 145-170. Mouton, The Hague.

Leach, B. F., 1979. Fish and crayfish from the Washpool midden site, New Zealand: their use in determining season of occupation and prehistoric fishing methods J. Archaeol. Sci. 6, 109-126.

Mellars, P., 1978. Excavation and economic analysis of Mesolithic shell middens on the Island of Oronsay (Inner Hebrides). In P. Mellars (ed.), The Early Postglacial Settlement of Northern Europe, 371-396. Duckworth, London.

Mellars, P. & Wilkinson, M., forthcoming. Fish otoliths as an indicator of seasonal occupation in Mesolithic shell middens.

Mikkelsen, E., 1978. Seasonality and Mesolithic adaptation in Norway. In K. Kristiansen & C. Paludan-Muller (eds.), New Directions in Scandinavian Archaeology Stud. Scand. Prehist. Early Hist. 1, 79-119.

Nikolsky, G. V., 1964. The Ecology of Fishes Academic Press, London.

Odum, E. P., 1975. Ecology 2nd ed. Holt, Rinehart & Winston, London.

Paine, R., 1957. Coast Lapp society; 1. Tromso Mus. Skr. 4.

Poiner, G., 1976. The process of the year among aborigines of the central and south coast of New South Wales Archaeol. Phys. Anthropol. Oceania 11, 186-206.

Reinsch, H. H., 1976. Köhler und Steinköhler. Ziemsen, Wittenberg.

Schalk, R. F., 1977. The structure of an anadromous fish resource. In L. R. Binford (ed.), For Theory Building in Archaeology, 207-249. Academic Press, London.

Smith, B. D., 1976. 'Twitching': a minor ailment affecting human palaeo-ecological research. In C. E. Cleland (ed.), Cultural Change and Continuity, 275-292. Academic Press, London.

Steinmetz, B., 1974. Scale reading and back-calculation of bream Abramis brama L and rudd Scardinius erythropthalmus L. In T. B. Bagenal (ed.), 148-157. The Ageing of Fish, Unwin, Old Woking.

Trout, G. C., 1954. Otolith growth of the Barents Sea cod Rapp. P-V. Reun. 136, 89-102.

Weatherley, A. H. & Rodgers, S. C., 1978. Some aspects of age and growth. In S. B. Gerking (ed.) Ecology of Freshwater Fish Production, 52-74. Blackwell, Oxford.

Wheeler, A., 1969. The Fishes of the British Isles and North-west Europe. Macmillan, London.

Wheeler, A., 1978. Problems of identification and interpretation of archaeological fish remains. In D. R. Brothwell, K. D. Thomas & J. Clutton-Brock (eds.), Research Problems in Zooarchaeology. Inst. Archaeol. Occas. Publ. 3, 69-75.

Williams, T. & Bedford, B. C., 1974. The use of otoliths for age determination. In T. B. Bagenal (ed.), 114-123. .The Ageing of Fish, Unwin, Old Woking.

Wollabaek, A., 1900. Some biological notes. In J. Hjort & K. Dahl (eds.), Fishing experiments in Norwegian fjords. Rep. Norw. Fish. Mar. Invest. 1, 96-119.

THE RELEVANCE OF BIRDS FROM COASTAL AND ISLAND SITES

By Don Brothwell, Don Bramwell and Graham Cowles

In 1848, the Danish zoologist, Japetus Steenstrup, was able to demonstrate on bird and other bone remains from Jutland shellmounds that a range of information on climatic conditions, vegetational cover, and hunting techniques could be derived from them. A hundred years later, Clarke's review on fowling in prehistoric Europe really indicated the failure of archaeology to make full use of this type of data on avifauna. A more recent review by Dawson (1969) shows that the scene is at last changing, and quite considerable effort is now directed towards bird remains in some parts of the world. Only a part of the material is of course concerned with sea and shore birds.

Usually only bones or bone fragments are available for study, although even pieces of egg shell are now being recognised, collected and studied from various sites. Feathers are generally only preserved in arid areas, for instance Peruvian coastal sites such as Ancon have produced fragments and even whole feather decorations. But the unexpected can happen in Northern Europe, for instance, the preservation of peacock feathers in the Gokstad Viking ship is reported by Brønstead (1940). Perhaps the ultimate in the use of bird remains was demonstrated by Kubler (1948) in his use of guano to evaluate time—a sort of 'coprochronology'.

As far as the excavation of bird bones in concerned, there appears to have been no evaluation so far of the possible differential preservation and survival of such bone, although it is often more fragile than mammal bone of comparable size. In the case of smaller species, recovery by careful sieving is demanded, but has not happened much in the past. Regarding the excavation of dune deposits, Dawson (1958) mentions a problem he has experience of, namely the misleading association of early and later bones as a result of gradual sand movement. This seems to be a particular problem at some New Zealand Moa sites.

The laboratory identification and analysis of bird bones is often no easy task. There are some 9000 avain species today, and although they obviously don't all pile into the same locality, quite a number of species may occur together, and may show confusing skeletal similarities. Some idea of the numbers of species which can occur relative to the number of bird bones in the site sample, can be seen in the following table of nine coastal and island sites, mainly from Britain and ranging from Pleistocene to Medieval in date (Table 1). In smaller samples, the species-to-bone ratio can clearly be as low as 1:3, and in bone samples of between 100 and 500, over 40 species can at times occur. Although identification may be difficult, and at times highly tentative, major errors are not so likely. To give a few examples of the usual kind of problem, Eastham (1968) had to consider the subtle differences between two

Table 1 Coast, island and estuarine bird species relative to bone numbers

Site	No. bones	No. species	Ratio S:B
1. Gorham's Cave, Gibraltar (Pleistocene)	500	45	1:11
2. Portchester Castle, U. K. (Roman)	378	16	1:24
3. Morton, Fife, U. K. (Mesolithic)	34	11	1:3
4. Ubeidiya, Israel (Pleistocene)	386	41	1:9
5. King's Lynn, U. K. (Medieval)	1970	31	1:63
6. King's Cave, Dura, U. K. (Medieval)	90	23	1:4
7. Newark Bay, Orkney, U. K. (Norse-Post-Medieval)	132	42	1:3
8. Great Yarmouth, U. K.	145	8	1:8
9. Southampton (Hamwih) U. K. (Saxon)	1023	16	1:64
10. Baynard's Castle, London (Medieval)	4522	60	1:75

Data from: (1) Eastham, 1968 (2) Eastham, 1975 (3) Cowles in Coles, 1971 (4) Tchernov, 1968 (5) Bramwell, 1977 (6) Cowles, unpublished (7) Bramwell, unpublished (8) Jones, unpublished (9) Coy, unpublished (10) Bramwell, 1975.

Mediterranean Manx Shearwater varieties, either of which could have occurred in Pleistocene bones from Gibraltar. In the case of fish-eagle remains from the Chatham Islands, transfer from the genus Haliacetus to the genus Ichthyophaga was undertaken on the evidence of the position of the outer proximal foramen of the tarsometatarsus (Harrison and Walker, 1973). The occurrence of immature bones may further complicate identification, as Bourne (1967) found in his study of subfossil petrel bones from the Chatham Islands.

As far as the more demographic types of information are concerned, avian studies are sadly behind the standards set for mammal work in archaeology, although probably the relevance of this kind of information on birds will never be so great anyway. Regarding numbers, in his study of bird material from King's Lynn, Bramwell (1977) admitted that although on the basis of the most frequently occurring bones, a minimum number of birds would be 327 individuals, all possible factors (including probable contemporaneity and spacial separation) suggested a figure approaching a thousand. This becomes a far more difficult exercise if various site levels are involved. For instance, bird bones were scattered through eight levels at King's Cave, Isle of Jura, but although a minimum number of 28 individuals could be estimated, considerably more may be indicated by this scattered assemblage (Table 2). Sexing produces more problems and again its relevance to zooarchaeology has still to be evaluated.

Coast and island bird geography during phases of the Pleistocene was certainly not the same as seen at present. Changes in climate and coastal ecology, together with the growing influence of Man by historic times could well have produced marked changes in avian distributions and densities. Marine conditions, including the nature and extent of inshore ice, and sea levels in relation to cliff sites suitable for nesting were clearly important variables in some areas. Regions, such as the Mediterranean, although cooler at times, may well have been better supplied with bird food and must have generally had more undisturbed breeding sites. Two fossil avifaunas are revealing from this point of view; the last Glaciation birds of the Grotte di Romanelli (containing 6 species of goose, 2 species of duck and a swan), and the Maltese sample from Har Dalaam (with 4 species of swan, 4 of goose, and 6 other Anatidae). Even if only winter visitors, they demonstrate that different ecological conditions, with lowered sea-level and greater rainfall probably occurred. Moreau (1954) acknowledges that some of the avifaunas are "baffling mixtures" and gives the case of a Magdalenian fauna from the Dordogne (Astre, 1950) in which the cold-adapted Snowy Owl Nyctea occurs in such numbers as to suggest that it was not merely a winter visitor; and yet these are associated with the Jay, Rook and Pheasant, birds suggesting very different habitats.

Considering another Pleistocene site, Devil's Tower in Gibraltar (Bate, 1928), it is notable that the 31 bird species are nearly all locally known today, suggesting that although population changes must have occurred, there was an overall persistence of genera in this area, perhaps over much of the Pleistocene. But Great Auk bones both at Devil's Tower and La Cotte de St. Brelade suggest that the Neanderthalers were already busy assisting at least this species towards extinction. Similar to Devil's Tower are the Gorhams Cave birds, with a few minor differences perhaps related to local topography;

Table 2 Occurrence of species in the various levels and trenches at King's Cave, Isle of Jura, Scotland

	Cave Levels								No. of individuals represented
	1	2	3	4	5	6	7	8	
Great Northern Diver (*Gavia immer*)	-	F	-	-	C	-	-	-	1
Fulmar (*Fulmarus glacialis*)	-	-	-	-	-	-	-	B	1
Shag (*Phalacrocorax aristotalis*)	-	D	-	B	-	C*	-	B*C	2
?Greylag Goose (*Anser anser*)	-	E	-	-	-	-	-	-	1
Teal (*Anas crecca*)	-	E	-	-	-	-	-	-	1
?Pochard(*Aythya ferina*)	-	-	-	-	-	-	-	B	1
White-tailed Eagle (*Hailiacetus albicilla*)	-	-	C	-	-	-	-	-	1
?Black Grouse (*Lyryrys tetrix*)	D	-	-	-	-	-	-	-	1
Black Grouse (*Lyrurus tetrix*)	F	F	-	-	-	-	-	-	1
Ptarmigan (*Lagopus mutus*)	-	-	-	-	-	-	C	BC	1
Red Grouse (*Lagopus l. scoticus*)	-	F	-	-	-	-	-	-	1
Crane (*Grus*)	-	C	-	-	-	-	-	-	1
Oystercatcher (*Haematopus ostralegus*)	-	-	-	-	-	C	BC*	-	2
Herring or Lesser Black-backed Gull (*Larus argentatus* or *L. fuscus*	E*	-	-	C*	-	C	C	C	2
?Herring Gull (*Larus argentatus*)	-	-	E	-	-	-	-	-	
Great Black-backed Gull (*Larus marinus*)	D	-	-	-	-	-	-	-	1
?Great Black-backed Gull (*Larus marinus*)	-	-	-	-	-	-	-	B	
Kittiwake (*Rissa tridactyla*)	-	-	-	-	-	-	C	-	1
?Kittiwake (*Rissa tridactyla*)	-	-	-	-	-	-	C	-	
Common or Roseate Tern (*Sterna hirunda* or *S. dougallii*)	-	-	-	-	-	-	-	-	1
Razorbill (*Alca torda*)	-	-	-	-	-	-	B	-	1
Guillemot (*Uria salge*)	-	-	-	-	B*	-	B*C	C	2
Puffin (*Fratercula arotica*)	-	-	-	-	-	-	B	C	1
Rock Dove (*Columba livia*)	F	-	-	-	-	-	-	-	1

Table 2 (contd.)

	Cave Levels								No. of individuals represented
	1	2	3	4	5	6	7	8	
Carrion Crow (*Corvus corone*)	F*	F	–	–	C	–	C	C	2
Hooded Crow (*Corvus corone cornix*)	–	–	–	F	–	–	–	–	1
?Wheatear (*Oenanthe oenanthe*)	–	–	–	–	–	–	B	–	1
									28 min. no.

Key: B–F = trenches
* = evidence of separate individuals

for instance, Devil's Tower seems to have been near perches suitable for the Guillemot, whereas colonies of Little Auk and Puffin may have lived near Gorham's Cave. This cave also had species of waterfowl and divers not often seen in the Mediterranean today. Clearly, the more bird material can be studied in this kind of detail, the more it may assist in general environmental reconstructions in relation to Man, but caution is needed in any detailed reconstruction of the distribution of individual species. The reason for this is that individual species can at times show rapid changes (as yet not well explained) in the extent of habitat preferences. The perfect example is the fulmar in Britain, which in 1878 had just extended its breeding range from St. Kilda to the Shetlands. A hundred years later, its breeding empire (admittedly of variable density) occupies more than half the coastline of Britain and Ireland (Fisher, 1946).

When considering a time-span for bird populations extending into the Pleistocene, there is clearly a need to keep in mind the possibility of adaptive micro-evolutionary change. In this respect, Tchernov (1968) has considered the Calandra Lark (Meanocorypha calandra) through time in the Near East, and and demonstrated slight size and morphological differences probably in response to gradually changing climatic conditions. There is also the difficult problem of to what extent birds have adapted to different habitats through time, a question critical to reliable environmental interpretation.

Regarding bird samples of Neolithic and more recent date, far more are available, but at the same time, there is the greater danger that human populations have had a detrimental or at least modifying effect upon these communities. In these samples, numbers of specimens representing a species may be very small, but occasionally may be notably higher. This is seen in bones from Dutch sites (Fig. 1) where A, B and C are Neolithic to Bronze Age sites and the rest Roman to Medieval—and mainly of coastal regions (Clason, 1967). Peak frequencies occur with Mallards (number 10), domestic goose (17), Barnacle Goose (23), White-tailed Eagle (27) and domestic fowl (28). Our general impression is that British sites tend to have more species present, and the common birds, excluding the domestic forms, are different. Iron Age birds from a stalled cairn on Eday, in Orkney, included at least 28 cormorants, still a common Orkney species, together with lesser numbers of Herring Gull, Great Black-headed Gull, Shag, White-tailed Eagle and Merlin. In the Jarlshof excavations, Gannet was most common (Platt 1934), although it then declined in numbers in Shetland and recolonised again only recently (Bourne and Dixon, 1974).

At the Mesolithic Morton site in Fife, Guillemot bones were most numerous (Cowles, 1971) and at Newark Bay, Orkney, Guillemot and Rock Dove—although the latter may have been domesticated by post-Norse times. This last site is noteworthy in having the largest number of species identified relative to the number of bird bones found (42 sp in 132 bones). As can be seen in the species list (Table 3), a possible Norse example of a Tawny Owl occurs, but is unknown in the islands today—preferring woodland cover usually. Avian support for some tree cover having occurred in the Northern Isles is also provided from Jarlshof in Shetland, where Magpie and Black Grouse demand at least a scatter of light woodland (Platt, 1956). Although food birds could have been imported, it seems unlikely that the other species can be so explained.

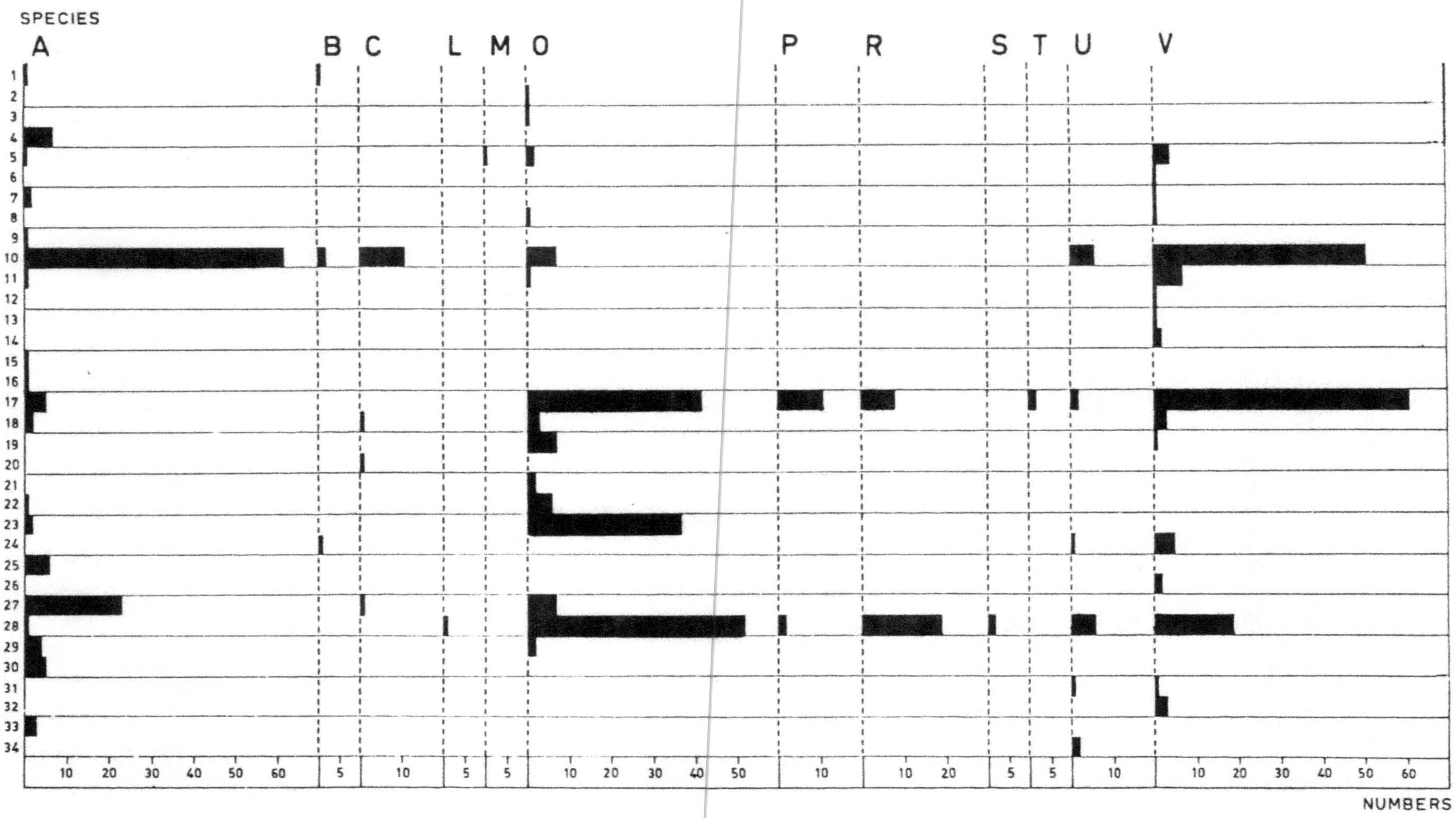

Fig. 1 The number of bird bones found at various Dutch sites, from Neolithic to medieval date. After Clason (1967).

Table 3 The birds identified at Newark Bay, Orkney

Species	Norse	Post-Norse	Totals
Lagopus lagopus (Grouse)	1	1	2
Pluvialis apricaria (Golden Plover)	4	4	8
?Limosa lapponica (Bar-tailed Godwit)	1	-	1
Numerius phaeopus (Whimbrel)	1	1	2
Gallinago gallinago (Snipe)	1	1	2
?Phalaropus sp. (Phalarope)	1	-	1
?Anas crecca (or querquedula) (Teal or Garganey)	2	-	2
Pinguinus impennis (Great Auk)	1	-	1
?Cepphus grylle (Black Guillemot)	2	-	2
Columba livia (Rock Dove)	2	9	11
Strix aluco (Tawny Owl)	1	-	1
?Turdus torquatus (Ring Ousel)	1	-	1
Anser anser (Grey-lag Goose)	2	7	9
Pluvialis squaterola (Grey Plover)	1	1	2
?Sp. Small wader	1	-	1
Larus marinus (Great Black-backed Gull)	1	1	2
?Sp. indet. sea bird	1	-	1
Sturnus vulgaris (Starling)	2	1	3
Corvus sp. (Crow or Rook)	2	-	2
Phalacrocorax carbo (Cormorant)	1	1	2
Phalacrocorax aristotelis (Shag)	1	2	3
Gallus sp. (Domestic fowl)	2	26	28
Numerius arquata (Curlew)	1	2	3
?Larus sp. (Small/Large Gull sp.)	1	1	2
?Larus or Rissa (? of Kittiwake)	1	-	1
Fratercula arctica (Puffin)	1	-	1
Puffinus puffinus (Manx Shearwater)	-	2	2
Anas platyrhynchose (Mallard)	-	5	5
Uria aalge (Common Guillemot)	-	7	7
Corvus corax (Raven)	-	1	1
Crex crex (Corncrake)	-	1	1
?Pluvialis sp. (Golden or Grey Plover)	-	1	1
Cepphus grylle (Black Guillemot or Puffin in one case)	-	2	2
Turdus merula (Blackbird)	-	1	1
Somateria mollissima (Eider Duck)	-	2	2
Alca torda (Razorbill)	-	1	1
Ardea cinerea (Heron)	-	1	1
Anser albifrons (White-fronted Goose)	-	1	1
Tadorna tadorna (Shelduck)	-	1	1
?Anas acuta (Pintail)	-	1	1
?Anus clypeata (Shoveler)	-	1	1
Aythya sp. (Pochard or Scaup)	-	1	1
Falco peregrinus (Peregrine Falcon)	-	1	1
Falco tinnunculus (Kestrel)	-	1	1
Lagopus lagopus (Grouse)	-	5	5
Scolopacidae sp. (2 species- ?one is Sanderling)	-	2	2
			132

As far as other parts of the world are concerned, Bourne (1967) has studied large numbers of Petrel bones from the Chatham Islands and concludes that they are not only an indication of early European slaughter of this avian group for food, but they also show that this island group was formerly a major breeding territory for petrels. In New Zealand, there is still uncertainty as to the extent evironmental change or the impact of Man, has changed the distribution of a large rail, the Takahe (Notornis), now restricted to small colonies in the South Island, but previously widespread. Dawson (1969) points out that (if Man is not to blame) we have two alternative possible explanations; either the bird changed its tolerance to certain environments, or there has been a contraction of its present type of habitat and climate in mountainous, snow-grass country.

Now a few words on the possible links in the past between human and coastal bird communities. Perhaps we could begin by mentioning a classic example, well known, but perhaps worth telling again. Probably at no time in the history of human settlement in the island group of St. Kilda, have population numbers gone much beyond 100 individuals. This small community, however, is an excellent example of the destructive reliance which can be put by humans on coastal birds, and of the serious impact we can have on other species. When Martin visited St. Kilda in 1679, vast numbers of gannets were breeding in the area and he estimated (probably overestimated) that more than 22,000 gannets were caught just in the previous year. In 1786, it was recorded that 1,200 gannets were taken in a single night. With the inevitable decline in this species, fulmars took over as the main bird of economic importance and by 1920, the annual catch of this bird was 9,600 (with puffins and a few other species also suffering considerable decimation). But in true Old Testament fashion, the fulmar had its revenge on St. Kildans for many years. By 1756, a serious disease was striking hard at newborn infants in the community, oddly enough, with twice the mortality in males. It is now a classic story in medical history that the reason for this condition, Tetanus infantus, was the application to the severed umbilical cord of a mixture, the major constituent of which was fulmar oil (Thompson, 1970). The curse of the fulmar remained until this century.

Of course, this is an extreme example of the impact of man on bird species, but there is no doubt that human communities have increasingly changed bird population sizes and ranges of breeding territory since Palaeolithic times. It is small consolation that the Greek Dionysius describes how attached gulls are to men (Pollard 1977), although there is no doubt that such an example of beneficial scavenging has encouraged their numbers at coastal ports and towns in recent centuries, as well as some land species at inland urban centres. But the facts are mainly against man. In the case of the Great Auk, now known from many European sites, numbers may concievably have been on the decrease even by Mousterian times in the Mediterranean. In Britain, the White-tailed or Sea Eagle occurred at a number of sites, but is now extinct as a breeding species. According to old accounts, although fast and fierce enough in the air, it could be killed by throwing stick or stone while it was on the ground too gorged with food to rise quickly into the air.

Island birds seem to have been particularly hard hit once human settlement was established. Rodriguez, an island in the Indian Ocean, has seen a serious decimation of sea birds in a few hundred years, by humans, rats, cats and forest fire (Bourne, 1968). Ascension Island, within its five centuries of human history, has seen a clearly major breeding area for sea-birds seriously threatened by Man, to the extent that the composition of sea-bird populations over a vast extent of ocean was probably dramatically affected (Olson, 1977). This point should perhaps be emphasised, that the repercussions of human interference (exploitation if your prefer it) can—and probably did in the past—go well beyond the immediate vicinity of human settlement.

In considering the exploitation for food of these birds, there is incidentally a danger that we will assess them according to the values of our own society today. In reporting on the occurrence of the Great Northern Diver at Portchester Castle, Anne Eastham (1975) remarks that the flesh is 'extremely unpalatable', but taste is unlikely to have been a deterrent in most earlier societies. Fenton (1976) notes that even in the nineteenth century, people in the Northern Isles would bury a cormorant for 24 hours to tenderise it and remove some of the fishy taste. Admittedly, eagles and falcons may have been killed more for the value of their feathers in fletching arrows and so on, but it would be revealing to look very carefully at even these remains, to see whether there is evidence that they also finished up in the cooking pot.

Seasonality is a subject which is intriguing to most archaeologists, although it does little more than confirm the basic fact that food availability is likely to show some fluctuation during the year, and that human communities were quick to realise that some potential food resources were linked to climatic or physiological cycles, and should be gone after or watched for at certain times of the year. Nevertheless, we would acknowledge that bird migration provides good opportunity to explore one aspect of seasonality in relation to coasts. In his study of bird remains from a shellmound in San Francisco Bay, Howard (1929) was able to demonstrate the occupation of the area throughout the year on the evidence of particular water birds. Species of ducks and geese indicated occupation from October to April, and the presence of abundant cormorant bones (nearly 50% being young nestlings) showed that shore birds were also exploited in the June-July period. What this doesn't reveal, of course, is whether the Amerindian groups moved spasmodically along the coast, using the food resources at various points during the year. On the evidence of a number of marine birds, especially cormorants, at a South African midden site, and the large percentage of fledglings, Avery (1977) concluded that occupation of the site by early Khoisan peoples was probably strictly limited to between December and April. These examples could be greatly extended, but the point really is to ask whether further conclusions could be derived from the bones and this kind of information. Would it be possible, for instance, by histological study, to demonstrate that birds were being caught after the breeding season and during their moult? Also, there is a growing literature on the internal parasites of wild birds, and Warner (1968) comments on the significance of disease in even causing the extinction of endemic Hawaiian avifauna. So could it be that disease is an increased hazard at breeding and moulting periods or following migrations, and again could the bones carry evidence?

The overall impression one has of avian studies in archaeology, which we have tried to show, is that there is great potential, but that progress has been slow, beyond basic reporting. There is no doubt that birds will eventually contribute more significantly to environmental archaeology.

BIBLIOGRAPHY

Astre, G., 1950. Faune magdalénne et axilienne de Rochereil (Dordogne). *Bull. Soc. Hist. Nat. Toulouse*, *85*: 151-171.

Avery, G., 1977. Report on the marine bird remains from the Paternoster midden. *S. Afr. Archaeol. Bull.*, *32*: 74-76.

Bate, D. M., 1928. Excavations at a Mousterian rock shelter at Devil's Tower, Gibraltar. *J. Roy. Anthrop. Inst.*, *58*: 92-113.

Bourne, W. R. P., 1967. Subfossil petrel bones from the Chatham Islands. *Ibis*, *109*: 1-7.

Bourne, W. R. P., 1968. The birds of Rodriguez, Indian Ocean. *Ibis*, *110*: 338-344.

Bourne, W. R. P. and Dixon, T. J., 1974. The seabirds of the Shetlands, pp. 130-144. *In* R. Goodier (ed.), *The Natural Environment of Shetland*. Nature Conservancy Council, Edinburgh.

Bramwell, D., 1974. Bird remains from medieval London. *London Naturalist*, *54*: 15-20.

Bramwell, D., 1977. Bird bone, pp. 399-402. *In* H. Clarke & A. Carter, *Excavations in King's Lynn 1963-1970*. Society for Medieval Archaeology monograph No. 7.

Brønsted, J., 1940. *Danmarks Oldtid*, Copenhagen. Quoted by Dawson, 1969.

Clark, J. G. D., 1948. Fowling in prehistoric Europe. *Antiquity*, *22*: 116-130.

Clason, A. T., 1967. *Animal and Man in Holland's Past*. Wolters, Groningen.

Cowles, G., 1971. Birds, pp. 350-351. *In* J. M. Coles, The Early settlement of Scotland, excavations at Morton, Fife. *Proc. Prehist. Soc*. *37*.

Dawson, E. W., 1958. Re-discoveries of the New Zealand subfossil birds named by H. O. Forbes. *Ibis*, *100*: 232-237.

Dawson, E. W., 1969. Bird remains in archaeology, pp. 359-375. *In* D. Brothwell and E. Higgs (eds.), *Science in Archaeology*. Thames & Hudson, London.

Eastham, A., 1968. The avifauna of Gorham's Cave, Gibraltar. *Bull. Inst. Arch. London*, *7*: 37-42.

Eastham, A., 1975. The bird bones, pp. 409-416. *In* B. Cunliffe, *Excavations at Portchester Castle*, *Vol. I, Roman*. London, Society of Antiquaries.

Fenton, A., 1976. Traditional elements in the diet of the Northern Isles of Scotland. pp. 1-16. Reps. Second Internat. Symp. Ethnol. Food Res. Helsinki.

Fisher, J., 1946. Watching Birds. Peguin Books, London.

Harrison, C. J. C. and Walker, C. H., 1973. An undescribed extinct fish-eagle from the Chatham Islands. Ibis, 115: 274-277.

Holyoak, D. T., 1973. An undescribed extinct parrot from Mauritius. Ibis, 115: 417-418.

Howard, H., 1929. Univ. Cal. Publ. Zool., 32: 301-394. quoted Dawson, 1969.

Kubler, G., 1948. Towards absolute time: guano archaeology. In: A reappraisal of Peruvian archaeology. Soc. Amer. Archaeo. Mem., 4: 29-50.

Moreau, R. E., 1954. The main vicissitudes of the European avifauna since the Pliocene. Ibis, 96: 411-431.

Olson, S. L., 1977. Additional notes on subfossil bird remains from Ascension Island. Ibis, 119: 37-43.

Peirce, M. A., Cheke, A. S. and Cheke, R. A., 1977. A survey of blood parasites of birds in the Mascarene Islands, Indian Ocean.. Ibis, 119: 451-461.

Platt, M. I., 1934. Report on the animal bones from Jarlshof, Sumburgh, Shetland. Proc. Soc. Antiq. Scot., 8: 313-319.

Platt, M. I., 1956. Report on animal bones. In: J. R. C. Hamilton, Excavations at Jarlshof, Shetland, Ministry of Works, Archaeological Report No. 1, Edinburgh.

Pollard, J., 1977. Birds in Greek Life and Myth. Thames & Hudson, London.

Tchernov, E., 1968. A preliminary investigation of the birds in the Pleistocene deposits of 'Ubeidiya. Israel Academy of Sciences and Humanities, Jerusalem.

Thompson, F., 1970. St. Kilda and Other Hebridean Outliers. David & Charles, Newton Abbot.

Warner, R. E., 1968. The role of introduced diseases in the extinction of the endemic Hawaiian avifauna. Condor, 70: 101-120.

www.ingramcontent.com/pod-product-compliance
Lightning Source LLC
LaVergne TN
LVHW072253300926
849160LV00018B/1661